# Optimizing Small Multi-Rotor Unmanned Aircraft

Team Gryphon's X8 CAD design for the IMechE UAS Challenge 2016.

# Optimizing Small Multi-Rotor Unmanned Aircraft

## A Practical Design Guide

**Stephen D. Prior**

Faculty of Engineering and Physical Sciences, The University of Southampton, Southampton, UK

CRC Press is an imprint of the
Taylor & Francis Group, an informa business
A BALKEMA BOOK

*CRC Press/Balkema is an imprint of the Taylor & Francis Group, an informa business*


Typeset by Apex CoVantage, LLC



*Library of Congress Cataloging-in-Publication Data*
Names: Prior, Stephen D. (Mechanical engineer), author.
Title: Optimizing small multi-rotor unmanned aircraft : a practical design guide / Stephen Prior, the University of Southampton, UK.
Description: London ; Boca Raton : CRC Press/Balkema is an imprint of the Taylor & Francis Group, an Informa Business, [2019] | Includes bibliographical references and index.
Identifiers: LCCN 2018035139 (print) | LCCN 2018036464 (ebook) | ISBN 9780429428364 (ebook) | ISBN 9781138369887 (hbk : alk. paper)
Subjects: LCSH: Drone aircraft—Design and construction.
Classification: LCC TL685.35 (ebook) | LCC TL685.35 .P75 2019 (print) | DDC 629.133/39—dc23
LC record available at https://lccn.loc.gov/2018035139

Published by: CRC Press/Balkema
Schipholweg 107c, 2316 XC Leiden, The Netherlands
e-mail: Pub.NL@taylorandfrancis.com
www.crcpress.com – www.taylorandfrancis.com

ISBN: 978-1-138-36988-7 (hbk)
ISBN: 978-0-429-42836-4 (eBook)

**This book is dedicated to the memory of my father, Robert Charles Henry Prior (1928–2008).**

This book is dedicated to the memory of my father, Robert Charles Henry Pryce (1928–2009).

# Table of Contents

# Tables

# Figures

# Abbreviations

| | |
|---|---|
| BEC | Battery Elimination Circuit (can be part of an ESC or standalone). |
| BLDC | Brushless DC Motor; a type of 3 Phase synchronous motor (80–90% efficient). |
| CAA | Civil Aviation Authority, UK governing body for civilian air regulation. |
| Capacity | The capacity of a Li-Po battery (mAh), i.e. 10,000 mAh = 10 Ah. |
| C-Rating | A rating of the max (cont.) current draw, i.e. 10C = 100 A for a 10,000 mAh Li-Po. |
| DoD | Depth of Discharge of a Li-Po battery (usually set at 85% of the capacity). |
| ESC | Electronic Speed Controller; a very fast, high current switch (98% efficient). |
| FM | Figure of Merit (A hover efficiency metric used in Rotary-Wing aircraft). |
| FW | Fixed-Wing aircraft. |
| g/W | A pseudo-efficiency metric (Thrust (g) divided by Input Power (W)). |
| HFC | Hydrogen Fuel Cell (converts Hydrogen into electrical energy (600 Wh/kg)). |
| $I$ | Motor Current (A). |
| $I_0$ | No-Load Current (A) – non-zero due to bearing friction, hysteresis, etc. |
| $K_m$ | Motor Constant used to size DC Motors (Nm/W); ratio of Torque/Power. |
| $K_T$ | Motor Torque Constant (Nm/A). |
| $K_v$ | Motor Voltage (Velocity) Constant (RPM/V)(rad/s/V)(V/kRPM). [$K_v$ is sometimes referred to as $K_b$ or $k_e$ (Back EMF or Electrical Constant).] |

Li-Ion — Lithium Ion cylindrical battery (243 Wh/kg, but very low C-Rating of 1 C).

Li-Po — Lithium Polymer battery (extremely low internal resistance ~ 0.001 ohm). [The best Li-Po batteries have a specific energy of approx. 200 Wh/kg.]

Li-S — Lithium-Sulphur battery (Pioneered by OXIS Energy in the UK - 400 Wh/kg).

MDO — Multi-Objective Design Optimization.

MTOM — Maximum Take-Off Mass (kg), (including the power supply and payload mass).

$M$ — Mass (kg).

PDS — Preliminary Design Specification.

PWM — Pulse Width Modulation (1–2 ms @ 50 Hz).

Prop — Propeller (Mainly FW, often referred to as a Rotor in RW) (30–80% efficient).

$Q_m$ — Motor Torque (Nm).

RC — Radio Control, sometimes referred to as Remote Control.

$Re$ — Reynolds number (ratio of inertial forces to viscous forces) (dimensionless).

Rotor — Rotor (Used mainly in RW aircraft or Rotorcraft).

RW — Rotary-Wing Aircraft.

RX — Receiver (on-board the aircraft).

S-Rating — Li-Po voltage given in multiples of 3.7 V/cell (1S), i.e. 6S = 22.2 V.

SE — Specific Energy of a Li-Po battery (Wh/kg), i.e. ranging between 150 and 200 Wh/kg.

$T$ — Thrust (N), often quoted incorrectly on manufacturer websites in gram (g).

TX — Transmitter (handheld unit usually controlled directly by the pilot in command).

UA — Unmanned Aircraft; sometimes referred to as RPAS, UAV or Drone.

$V$ — Motor Voltage (Volt).

$W$ — Weight (N), 1 kg = 9.81 N (2 dp).

# General assumptions

1. Assume the take-off point is at a Thrust to Weight ratio of 1:1.
2. Assume the Li-Po battery to be flat at 85% of its rated Capacity (depth of discharge).
3. Assume that the manufacturer's motor/prop data is imprecise; always conduct your own tests.
4. If using co-axial rotors, assume a reduction in overall thrust of 15–20%, when compared to two isolated rotors.
5. Propeller (Rotor) efficiency increases with diameter, pitch, speed (to a point) and supply voltage.
6. A well-designed system will have a target (g/W) ratio of about 10:1 at the take-off point.

# Design constraints (some governed by the UK IMechE UAS challenge)

*Figure 0.1* The author's HALO Y6, winner of the DARPA UAVForge Competition 2012 (Prior *et al.*, 2013).

1. Design the propulsion system take-off point to be at the 50% throttle mark.
2. Design the system to carry a payload of up to 3 kg (IMechE UAS Challenge Rules).[1,2]

1 Note that in the IMechE (2017) UAS Challenge the payload requirement was replaced with three separate challenges. Available from: www.imeche.org/get-involved/young-members-network/auasc

2 The IMechE (2018) UAS Challenge has now dropped the endurance challenge mission. Available from: www.imeche.org/docs/default-source/1-oscar/uas-challenge/uas-challenge--competition-rules--issue-7-1.pdf?sfvrsn=2

3. Design the system to have a Thrust to Weight ratio within the range of 2:1 to 3:1.
4. Design the MTOM to be < 7 kg (CAA regulations/IMechE UAS Challenge Rules).
5. Design the system to use a 6S (22.2 V) Li-Po power supply (efficiency increases with voltage).
6. Design the system to be portable and occupy the minimum space possible when not in use.
7. Design the system to be frangible in line with CAA 722 guidelines.
8. Utilise system failsafe measures, such as link loss, low battery voltage, GPS loss, etc.
9. Design the system to use a Pixhawk flight controller – low cost, robust and reliable.

# Preface

This design guide was written to capture the author's practical experience of designing, building and testing multi-rotor systems over the past decade. The lack of one single source of useful information, meant that the last 10 years has been a steep learning curve, a lot of self-tuition and many trial and error tests. Lessons learnt the hard way are not always the best way to learn. As one of my previous lecturers said: 'Learn from the mistakes of others, rather than from your own mistakes'. However, when you are operating at the cutting edge, it is hard to find examples to follow.

Some aspects of this guide are deliberately focussed on design solutions to the IMechE UAS Challenge,[3] which is an international design competition set up by the Institution of Mechanical Engineers in the UK and first run in the summer of 2015. The basic premise of this competition is to design, build and test a small (< 7 kg), low-cost (< £ 1k), autonomous Unmanned Aircraft (UA) (FW or RW) to deliver a humanitarian aid payload (up to 3 kg)[4] accurately to a target GPS waypoint location approximately 1 km away from the take-off point. An extension to the 2017 competition rules involved the addition of an endurance mission and a reconnaissance mission.[5] In 2018, this was further amended to remove the endurance element.

## The IMechE UAS challenge competition (2015–2018)

An undergraduate team from the University of Southampton won the inaugural event in 2015 with a fixed-wing aircraft design. In 2016, our three rotary-wing aircraft designs were placed 2nd, 3rd and 4th respectively, winning

3 IMechE UAS Challenge (2016) Available from: www.imeche.org/news/news-article/institution-launches-unmanned-aircraft-systems-challenge-2016
4 Not an absolute limit, as one Southampton team carried 3.5 kg in the 2018 competition.
5 IMechE UAS Challenge (2017) Available from: www.imeche.org/docs/default-source/1-oscar/Get-involved/uas-challenge--competition-rules--issue-7-final-extended-deadline.pdf?sfvrsn=0

*Figure 0.2* Example platforms from the IMechE UAS Challenge (2015–2017).

prizes for Innovation, Autonomous Operations and Safety & Airworthiness. For the 2017 competition, we entered two teams of four MSc students. Both were rotary-wing solutions and were placed 2nd and 3rd overall, winning prizes for the Payload challenge and GPS Accuracy challenge.

In the 2018 competition, we entered one rotary-wing, Y6 and a tail-sitter design. The Y6 was placed 2nd overall and the tail-sitter won the Design award. To date, no rotary-wing platform has won the overall prize. It is my aim, as the overall team leader, to change this dynamic and prove that a rotary-wing solution can be better than a fixed-wing solution.

# Acknowledgements

Over the past 25 years, I have had the pleasure of teaching many bright and gifted students in the UK, Hong Kong and Taiwan. This has taught me to always strive for perfection and in doing so, together we have achieved some measure of success. Whether in the field of Automobile Engineering, Mechanical Engineering, Robotics, Product Design & Engineering or more recently Aeronautics, Astronautics and Computational Engineering, taking risks and challenging the status quo has always paid off.

This book is the product of 10 years of experiential learning, trial and error and yes, some spectacular failures, each of which taught a hard, but essential lesson in real-world dynamics. Over the last 3 years, the book has come together, slowly but surely, thanks in the main to my loving and understanding wife Prof. Siu-Tsen Shen. This period also coincided with the birth of our first son, Jaime L. Prior. I thank them both for teaching me humility, perseverance and the importance of family. To this end, I would also like to thank the many family and friends who have also supported me through good and bad times. Ma, Eamonn, John, Tony, Jimmy, Peter, Gus and Gina, you know who you are.

A special mention goes out to my research team in the Autonomous Systems Lab, who have now mostly graduated through the system or have progressed onto better things: Dr Siddharth Odedra, Dr Mehmet Ali Erbil, Dr Mantas Brazinskas, Witold Mielniczek, Chris Barlow, Darren Lewis, Dr Chang Liu and Ayodeji Opeyemi Abioye.

To all the many students and staff that I have worked with on projects over the years, at Southgate Technical College, Middlesex University, Hong Kong Tsing-Yi Technical College, National Formosa University and The University of Southampton, I thank you.

Several Southampton students, past and present, that have helped directly with data in this book deserve to be highlighted, these are Adrian Weishaeupl, Yehya Abdallah, Raam Sundhar Sampath, Kufre Etok, Colin Whiteley, Pratik Joshi, Stephen Mace, Josh Mills, Jon Arnarson, Mike Emptage, Josh

Davies, James Perrett, Martin Garcia and Achal Mittal. In terms of getting to The University of Southampton, I'd like to personally thank Prof. Andy Keane and Prof. Jim Scanlan; your dedication, intelligence and can-do attitude continue to inspire!

To the people that get things done; thanks go to Dr Mario Ferraro, Andrew Lock, Bob Entwistle, Prof. Keith Towell, Dr Sarvapalli Ramchurn and Dr Angelo Grubisic. External, but none the less important, Jim Gibson (Tethered Drone Systems Ltd), Dr Peter Saddington (Tekever Ltd), Prof. Peter Wilson (Bath), Dr Glyn Thomas, Prof. Rob Richardson (Leeds) and Dr William Crowther (Manchester).

Finally, to all the people and organisations that provided access to data, images and graphs, I thank you for making this book complete.

## Copyright notice

# 1 Introduction

The majority of RC component (Battery, ESC, BLDC Motor and Propeller) manufacturers do not state accurate empirical performance measurements for these vital parts. Those that do sometimes state inaccurate and/or inflated values, presumably to drive sales and to fool the unsuspecting consumer.

It is therefore vital for the user to seek corroboration of data online and/or conduct their own static thrust tests,[1] before specifying and purchasing a complete system solution. A rare exception to this situation is the data provided by the US-based propeller manufacturer APC (Advanced Precision Composites) who provide computerised performance data on their full range of 493 propellers used in both fixed-wing and rotary-wing (multi-rotor) aircraft.[2] However, unfortunately a recent comparison of UIUC wind tunnel data[3] has shown this to over-estimate the maximum efficiency by up to 14%, as can be seen in the figure below for the APC 11″ × 3.8″ SF propeller running at 3000 RPM. Apparently, this is due to the NASA TAIR code[4] that they use to predict the airfoil lift and drag. APC are looking into improving this code.

The majority of this design guide is focussed towards the optimum design of an all-electric multi-rotor configuration. However, many of the findings will be also applicable to small, fixed-wing aircraft.

Once a system configuration (Helicopter – single main rotor, tandem, co-axial), Bicopter, Tricopter, Quadrotor, Hexrotor, Y6, Octocopter, X8, etc.) has been chosen, it is important to draft a Preliminary Design Specification (PDS). This sets out the important parameters of the system design process.

1 Dr Kiwi (2018) *BLDC Motor/Prop Tests*. Available from: www.flybrushless.com/user/profile/14

2 APC (2018) *APC Prop Data*. Available from: www.apcprop.com/v/PERFILES_WEB/listDatafiles.asp

3 UIUC (2018) *Propeller Data Site*. Available from: http://m-selig.ae.illinois.edu/props/propDB.html

4 APC (2018) *Airfoil Design Data*. Available from: www.apcprop.com/technical-information/engineering/

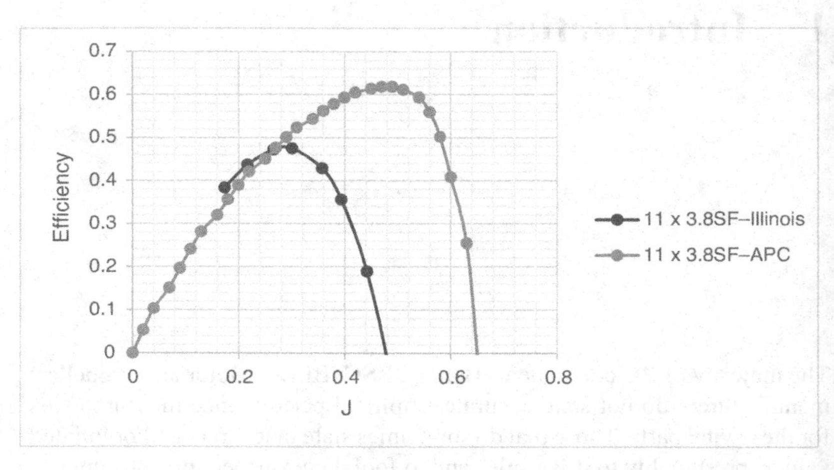


*Figure 1.1* Comparison of wind tunnel data with computerised data for an APC 11″ × 3.8″ SF propeller.

Given the design constraints of the IMechE UAS Challenge, two of the most important elements of the PDS, are the desired endurance (min) and the payload capacity (kg). A typical value for the maximum endurance of a small multi-rotor (< 7 kg) is 20 min; however, the best of class has an endurance of around 30–40 min (usually quoted without payload).

The ultimate goal should be to exceed the 1 hr endurance target, whilst carrying a payload of say, 5 kg, which would cover 90% of all current payloads. To achieve this, would require a design at an MTOM of approximately 15 kg, probably utilising a more exotic power source such as a Hydrogen Fuel Cell, Fuel Engine or Hybrid system.

In terms of payload capacity, most small multi-rotors (< 7 kg) are designed to carry a small payload of about 0.5–1 kg (max). Clearly, one of the goals of the IMechE UAS Challenge organisers was to push the boundaries of what is technically possible in terms of payload, endurance and autonomy.

Many commercial multi-rotors carry a small camera, which usually forms their payload. Improvements to these devices have allowed high definition (HD) cameras, with recording capability, to be incorporated into small, lightweight (< 30 g) and low-cost packages (< £ 30).[5]

5 HobbyKing (2018) *HD Camera*. Available from: www.hobbyking.com/hobbyking/store/__17200__HD_Wing_Camera_1280x720p_30fps_5MP_CMOS.html

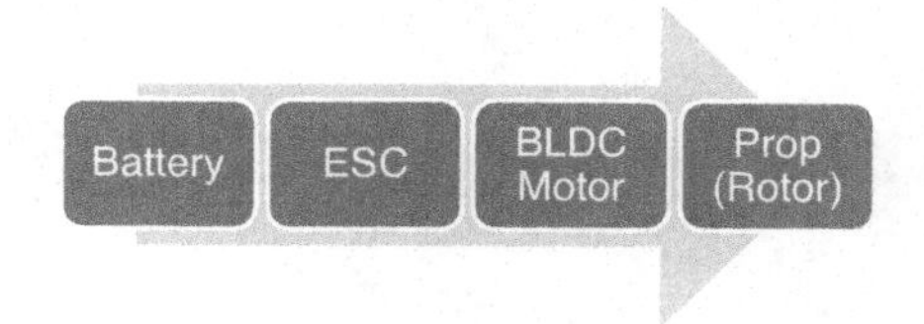


*Figure 1.2* The four main components of an RC electric aircraft propulsion system.

The aim of multi-rotor multi-objective design optimization (MDO) must therefore be to maximise the endurance, together with the payload capacity. By understanding the fundamental principles of each sub-system, it is possible to select the 'best in class' individual components. In the following chapters, each of the four main propulsion system components will be examined in detail, starting with the power supply and moving towards the propeller (rotor).

# 2 Lithium Polymer battery (power supply)

Lithium Polymer battery technology (with low internal resistance: typically 0.001 Ω) currently gives the highest Specific Energy (150–200 Wh/kg) of any type of rechargeable battery.[1] Given the design requirement to supply high current, 40–150 A (continuous) at low voltage DC (typically 7.4–22.2 V), the Li-Po is a natural choice (see Appendix A). Li-Po batteries typically range from 10 to 120 C.

***Tip 1:*** To calculate the max continuous current capability, multiply the 'C' rating by the capacity (Ah).

Rival systems such as Li-ion, Li-S, Hydrogen Fuel Cells (HFC), Super Capacitors and Photo Voltaic (PV) systems, etc., lack the high current draw capability, have refuelling/disposal issues, are large in size & mass and can be very expensive, making them impractical at this time. Having said that, if money is no barrier and you are willing to build a very lightweight UA, which is on the edge of its design envelope, you can achieve hover endurance for a multi-rotor in the region of 2–6 hr using these systems.[2,3,4,5]

- Li-Ion batteries (High SE (243 Wh/kg), very low (0.5–2) C-rating, 3.6 V/cell).

1 Cylindrical Panasonic cells (NCR18650B) have a 21% higher specific energy (243 Wh/kg).
2 Guinness World Records (2018) *Multicopter Flight Duration*. Available from: www.guinnessworldrecords.com/world-records/longest-rc-model-multicopter-flight-(duration)/
3 HUS (2018) *Hydrgoen Multi-rotor*. Available from: www.hus.sg/#!hydrogen-multi-rotor/ccm4
4 Energyor (2018) *Hydrogen Drone*. Available from: http://energyor.com/news
5 Powerlight Tech (2018) *Directed Energy Drone*. Available from: http://powerlighttech.com/

- Li-S batteries (High SE (400 Wh/kg), not commercially available yet, low C-rating, 2.1 V/cell).
- Hydrogen Fuel Cells (Very high SE (600 Wh/kg), high mass, large size, very expensive, pollution).
- Photo Voltaic (PV) – Solar (Low efficiency < 44%, large surface area (250 W/m$^2$), low current).[6]
- Super Capacitors (High specific power, low specific energy, low cost and low mass).
- Directed Energy (Laser Power) (Low efficiency < 20%, line of sight operation, safety concerns).
- Nuclear batteries (Extremely high SE, very low efficiency, large mass, safety concerns).

In this context, an IC engine cannot be beaten from a Specific Energy standpoint (900+ Wh/kg); however, the drawbacks of high mass, fuel safety, heat signature, noise, pollution and vibration make this choice undesirable. However, recent hybrid IC Engine/Li-Po battery solutions have been shown to work.

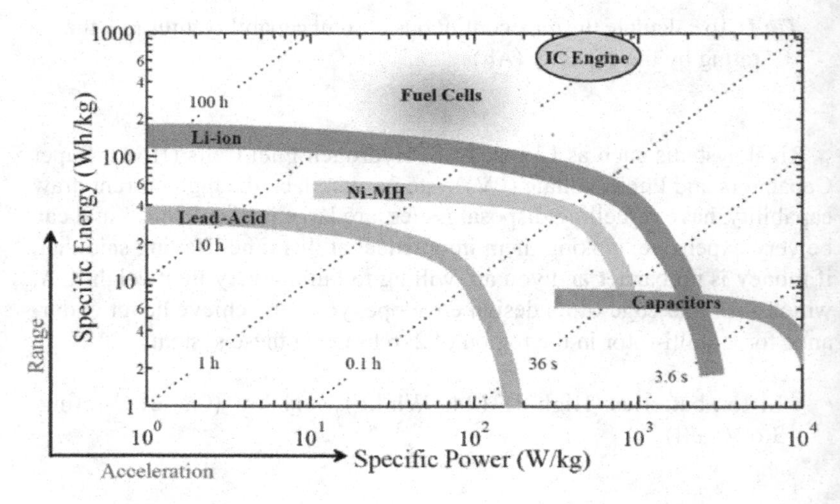


*Figure 2.1* Ragone plot of energy storage comparison in terms of performance (Srinivasan, 2008).

6 Alta Devices (2018) *Photo-Voltaic Small Unmanned Systems*. Available from: www.altadevices.com/unmanned-systems/

However, it is interesting to note that Petrol, as a fuel, has a Specific Energy of 12,200 Wh/kg; this puts all forms of rechargeable battery into perspective.[7] That said, electric motors are nearly four times more efficient than IC Engines, so the equation is more likely to reach equivalence at 3050 Wh/kg. Given the extra mass of an IC engine, it has been stated that this could come down to equivalence at about 340 Wh/kg. The growth of all-electric vehicles, such as the Tesla range, is testament to this.

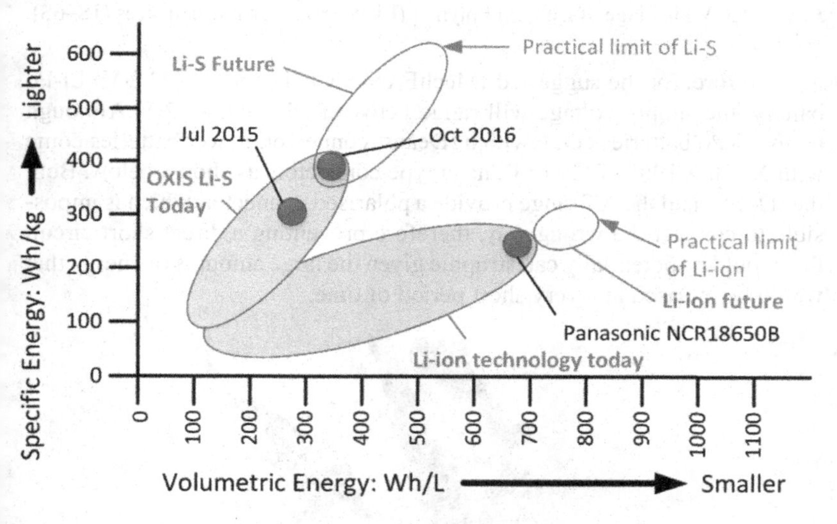


*Figure 2.2* Comparison of Li-S batteries with other rechargeable battery types (Oxis Energy, 2018).[8]

Li-Po batteries come in all shapes and sizes from very small – 250 mAh – 50 g (1 Cell/1S) to very large – 30,000 mAh – 3.67 kg (6 Cell/6S) and are often designated as 6S1P, 3S2P, etc. (this refers to the number of cells in Series (increased voltage) and the number of cells in Parallel (increased capacity)). Each cell of a Li-Po battery is nominally rated at 3.7 V. When fully charged this will reach 4.2 V/cell, and when depleted this will drop to about 3.2 V/cell (the range (charged~discharged) is approx. 1 V/cell).

7 Transtronics Inc. (2017) *Energy Density*. Available from: https://xtronics.com/wiki/Energy_density.html

8 Oxis Energy (2018) *Lithium-Sulfur Battery*. Available from: https://45uevg34gwlltnbsf2plyua1-wpengine.netdna-ssl.com/wp-content/uploads/2017/07/OXIS-Li-S-Long-Life-Cell-v4.03.pdf

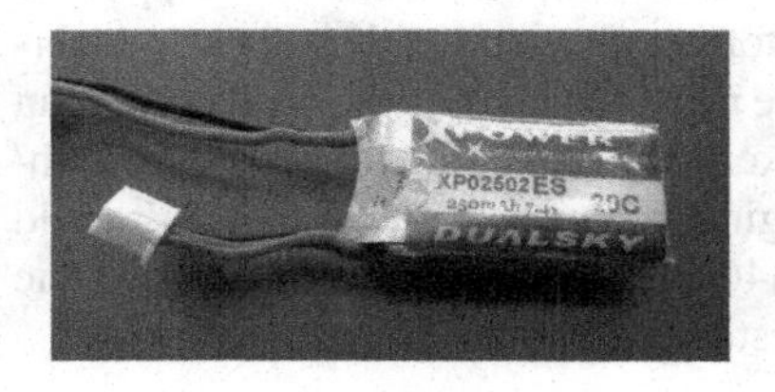

*Figure 2.3* Wide range of Lithium Polymer (Li-Po) rechargeable cell sizes (1S–6S).

Therefore, for the suggested IMechE UAS Challenge 6S (22.2 V) Li-Po battery, the supply voltage will range between 25.2 and 19.2 V. Although many Li-Po batteries come with a 'Deans' connector, newer batteries come with XT60/XT90/XT120 or Banana type connectors as shown below. Both the 'Deans' and the XT range provide a polarised connector, which is impossible to connect the wrong way, therefore preventing a direct short circuit that would be potentially catastrophic given the large amounts of energy that would be released in a very short period of time.

*Figure 2.4* XT, Deans and Banana main battery polarised connectors.

**Example 1**

The MaxAmps 4S (14.8 V), 10,900 mAh, 120 C, Li-Po battery[9] has a mass of 919 g, therefore, it has a Specific Energy of (14.8 x 10.9)/0.919 = 175.5 Wh/kg. This battery is therefore theoretically capable of supplying a massive 1308 A (120C continuous) and can be charged at 5C (54.5 A). The cost (as at June 2018) is US$ 310 (£ 230), which is expensive, however, the manufacturer claims 500 charge/discharge cycles before replacement, so this may be seen as good value for money.

9 MaxAmps (2017) *Lithium-Polymer Batteries*. Available from: www.maxamps.com/lipo-10900-4s-14-8v-battery-pack

***Tip 2:*** To calculate the SE (Wh/kg) multiply the voltage (V) by the capacity (Ah) and divide by the mass (kg).

## 2.1 Safety considerations

Li-Po batteries should only be charged using an appropriate Li-Po charger (with the battery stored in a charging bag) and should never be discharged below their minimum voltage of about 3.2 V/cell. Failure to adhere to this advice could create a thermal runaway (over 150°C), resulting in a Lithium fire. This type of fire can only be extinguished using a suitable fire extinguisher (see Appendix B). Never use a Li-Po battery above its C-Rating (Continuous or Intermittent) or if the battery pack is swollen, as this is a sign of damage. A low-cost Li-Po voltage alarm and cell tester (shown below), which shows the individual cell voltage, as well as the total voltage, is an excellent purchase and may prevent damage to the battery and the unmanned aircraft. Note: High-quality Li-Po battery chargers balance the voltage of every cell.

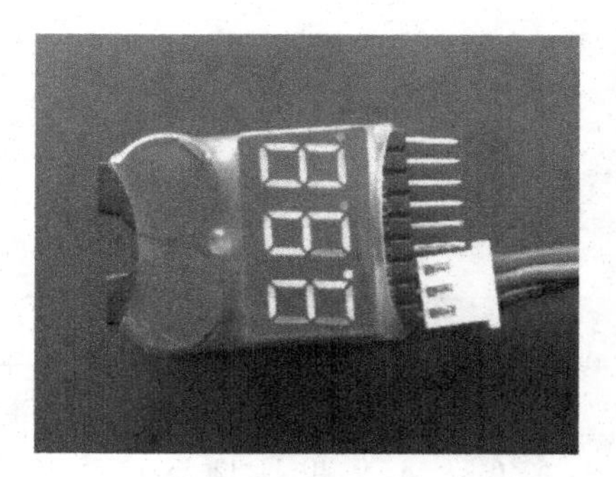

*Figure 2.5* A low-cost Li-Po voltage alarm and cell tester.

## 2.2 Lithium Polymer discharge characteristics

The discharge curve of a typical Li-Po battery is affected by the discharge current (A) and operating temperature (°C), the harder you work it, the quicker it depletes. Li-Po batteries perform better in warmer, rather than colder, conditions.

As a general rule of thumb, exceeding the stated 'C' rating of a Li-Po battery will damage one or more of the cells and will eventually result in premature failure to hold charge. Likewise, operation in extremely cold conditions will reduce the operational life of the battery. Some UAV operators use heating elements to warm the battery prior to take-off, and sometimes during the flight.

Note the sharp cut-off point at the RHS of the graph (highlighted area) (see below) where the cell voltage drops off rapidly. This is why it is recommended not to operate outside the 85% (max) cut-off point, to enable enough power to land safely under any given environmental conditions.

> ***Tip 3:*** Never calculate your endurance using 100% of the battery capacity, I recommend using an 85% DoD.

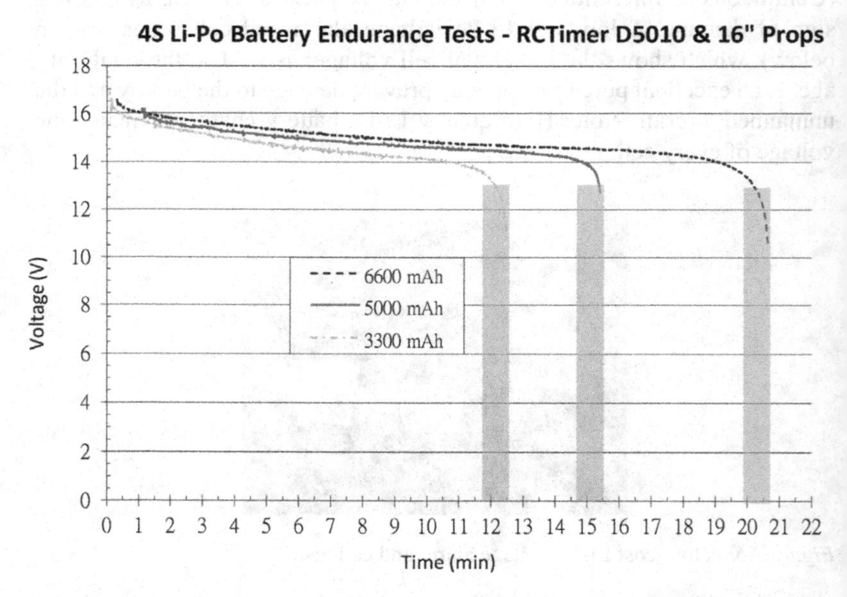


*Figure 2.6* Typical discharge 'S' curves of Lithium Polymer rechargeable batteries.

# 3 Electronic Speed Controller (ESC)

The Electronic Speed Controller (ESC) is a very high efficiency (98%), low resistance (typically 0.01 Ω), high current switch. This component takes the control input (PWM throttle signal) from the Autopilot via the TX/RX circuit and converts this to a three-phase signal which can be interpreted by the BLDC motor to control the speed of the motor output shaft (prop). Many ESCs also incorporate a Battery Elimination Circuit (BEC) to regulate a voltage (5 V) to power the receiver or another device, removing the need for a separate Li-Po battery for this purpose, however, a word of caution; they are limited to a few Amps at most, so be careful not to overload them by asking them to drive large servos!

Most ESCs operate over a wide range of input voltages (2–6S) and come in various current capacities which are clearly marked on the outer cover. They are generally lightweight and low cost; however, they are prone to overheating, which can result in a thermal shutdown causing a multi-rotor crash scenario, if pushed too hard or incorrectly specified at installation.

> ***Tip 4:*** Always ensure that your ESCs are mounted in a position where they can be naturally ventilated.

The input frequency (600 Hz) can also be marked on the cover. It is claimed that higher frequency ESCs are better at controlling the speed of large (inertia) BLDC motors/props used in big multi-rotor designs. Some bigger platforms operate at up to 14S, but require special ESCs.

*Figure 3.1* A typical Electronic Speed Controller (2S–4S) 30 A.

All ESCs have two wire inputs (+ve and –ve direct from the Li-Po battery via a power distribution board) and three wire outputs (A, B and C phases of the BLDC motor). There is also the RX (three small wires – signal (orange/yellow/white), +ve (red) and –ve (black/brown)) on the input side.

Avoid long cable runs (> 25 cm), especially on the power side. If this is impossible, add suitable extra capacitors in-line with the ones already mounted on the ESC. If you are conducting tests on a static setup, it is useful to deploy a servo tester connected to the RX cable to easily adjust the BLDC motor speed (see below).

*Figure 3.2* A servo tester which mimics the RC controller PWM signal to control a BLDC motor.

This unit is basically a PWM emulator, which replaces the RX output, and controls the motor speed. The pulses range from 1 to 2 ms, with 1 ms (off), 1.5 ms (half speed) and 2 ms (full speed) at 50 Hz. Therefore, the signal refreshes every 20 ms. In practice, these tend to be set at 1060 μs (off) and 1860 μs (full speed) to be compatible with all systems.

Simon Kirby's Open source firmware (SimonK) ESCs, which have been specifically designed and tuned for multi-rotors (as opposed to standard setups) have recently found favour amongst RC enthusiasts who claim that they provide a faster response and smoother performance with greater efficiency.[1]

As a general rule of thumb, select an ESC which has about 50% more rated current capacity than that required from your system, i.e. if your motor runs at 16 A max, then choose a 25 A rated ESC.

> ***Tip 5:*** If your individual motor is likely to draw 16 A, choose an ESC with 50% more capacity, i.e. 25 A.

ESCs have many other built-in functions, most of which were designed for helicopter or fixed-wing aircraft use. However, the designer of a multi-rotor system should be aware of these and what they all do. Failure to understand these can lead to catastrophic accidents.

In the old days, these functions were set manually using the bleep-bleep-bleep sequence of the RC controller and motor (a mind-numbing experience), hence the introduction of the programming card (see Fig. 3.3). The latest innovation in ESCs means that these can now be set-up in software, thus avoiding all the hassle and expense.

Most of these settings default to what is common amongst fixed-wing and rotary-wing setups, i.e. battery type, music (Li-Po cells), timing mode, however, a few are specific to helicopter and fixed-wing setups such as brake, start mode and governor mode; these should generally be either turned off or set to normal for multi-rotor use. On set-ups with very large props, you might want a very soft start mode.

## 3.1 Timing

One ESC setting which is not well understood is the timing setting. The purpose of this is to decide when to switch to the next sequence of energising the BLDC motor coils.

This usually has three (sometimes four) possible settings, Low, Middle (Default, sometimes called Automatic) and High. Low pole count motors should be set to Low, High pole count motors should be set to High and all others or BLDC motors where the pole count is unknown, should be set to Middle (Default or Automatic).

1 Github (2018) *Software Repository*. Available from: https://github.com/sim-/tgy

On MayTech ESCs, Automatic determines the optimum motor timing. Low is between 7° and 22° (2–4 pole motors) and High is between 22° and 30° (6 poles or more motors). Generally, low timing is more efficient, whereas high timing provides more power and speed at the expense of efficiency.

The timing setting will also affect the $K_v$ value of the BLDC motor, which is discussed in detail in the next chapter. Basically, as the timing is increased, so the $K_v$ value also increases (by up to 8% in one study).[2]

To analyse the relationship between the timing setting and the motor speed constant, $K_v$ tests on two different BLDC motors were conducted, the results of which are shown in the table below.

*Table 3.1* Data showing the Motor Timing Influence on $K_v$ for two BLDC motors.

| *Motor Timing* | $K_v$ |
|---|---|
| Foxtech W61-35 manually spun | 315.07 |
| Foxtech W61-35 7°–22° Timing (Low) | 333.12 |
| Foxtech W61-35 22°–30° Timing (High) | 374.86 |
| Foxtech W61-35 manufacturer stated $K_v$ | 330 |
| T-Motor U5 manually spun | 361.75 |
| T-Motor U5 7°–22° Timing (Low) | 374.92 |
| T-Motor U5 22°–30° Timing (High) | 405.73 |
| T-Motor U5 manufacturer stated $K_v$ | 400 |

## 3.2 Cut-off voltage

The one setting which is vital to get right is the cut-off type and associated voltage. If this is set to cut off and the receiver loses the signal from the TX, then the motor will failsafe i.e. shut down with associated consequences! A far better arrangement may be to set this to soft cut and set the cut-off voltage to low or middle. If the TX fails, the multi-rotor should just move the throttle to low or middle (essentially a hover situation) which is fairly safe. Setting this to high could result in a flyaway scenario, so apply caution to this setting and bench test (without prop) before actual flight!

2 Myers, K. (2016) *Timing Test*. Available from: www.theampeer.org/timing/timing.htm

## 3.3 Switching frequency

Occasionally an ESC may allow the user to alter the switching frequency between the standard 8 kHz frequency and the higher 16 kHz frequency. This controls how quickly the on-board Field Effect Transistors (FETs) can switch on and off the voltage supply. Although 16 kHz is more efficient, this is not normally used due to the higher RF noise signature emitted.

Whilst we are discussing switching, a common misconception is that the ESC effectively reduces the supply voltage as the controller's PWM signal is changed. This is incorrect, the supply voltage remains unchanged, however, the time that this is effectively switched on varies between completely off (0 V), 50% on (11.1 V) and 100% on (TOW) 22.2 V for a 6S Li-Po battery setup.

> ***Tip 6:*** When in doubt, test these ESC programmable settings on the ground before flying for the first time.

Setting of the previously mentioned features used to be done, rather laboriously, via a sequence of ESC bleeps set via the RC controller. This was not only monotonous, but also open to errors.

An example of this procedure from YGE[3] is shown below:

> Timing (♪♪♪) Timing setting: move stick again into neutral position: The ESC starts with a single beep (30°) and proceeds up to 7 beeps (Auto timing). Example: To set 18°: Move the stick to full power at the third beep signal. ♪ 30° ♪♪ 24° ♪♪♪ 18° ♪♪♪♪ 12° ♪♪♪♪♪ 6° ♪♪♪♪♪♪ 0° ♪♪♪♪♪♪♪ Auto timing – At the desired beep count, move the stick to full power. Acknowledgement: ♪ ♪

These days, the introduction of the programming card (shown below), has enabled all the settings to be configured in a single operation, thus reducing costly errors.

3 Young Generation Electronics (2016) *Finest Brushless Controller*. Available from: www.yge.de/en/home-2/

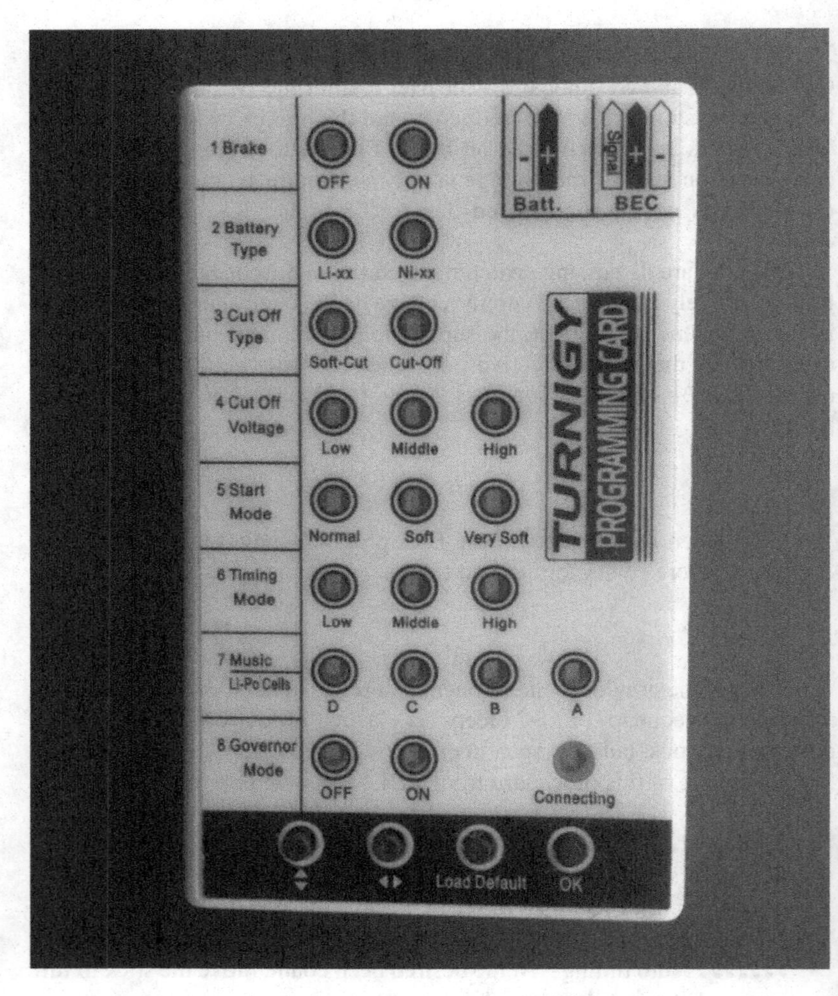


*Figure 3.3* A Turnigy programming card for setting ESC parameters.

# 4 The Brushless DC (BLDC) motor (outrunner)

The ubiquitous Outrunner BLDC motor has the rotor on the outside, incorporated into the casing. This is opposite to the older and now less used Inrunner. The BLDC motor is essentially a standard DC motor, without the old carbon brushes, which caused friction, sparks and ultimately failure due to wear. However, to complicate things further, there are three windings (A, B and C) and therefore this is classified as a three-phase (3P) BLDC motor, similar in nature to a synchronous AC motor.

The ESC described in the previous chapter, essentially takes the DC voltage input and outputs a three-phase AC trapezoidal (or sinusoidal) voltage to each of the three windings within the BLDC motor. The timing and length of these pulses is crucial to correctly control the speed and direction of the motor.

The current generation of ESCs use a rather crude scalar form of control, called six-step commutation, with current being passed through only two of the three windings at any one time. The latest innovation in the design of ESCs is by deploying a vector control strategy, called Field Oriented Control (FOC), which claims higher torque, less torque ripple and faster response rates.[1]

Sometimes a motor manufacturer may quote a motor designation such as 12N 14P, this refers to the design characteristic of 12 electro-magnets on the stator teeth (always a multiple of three) and 14 permanent magnets in the rotor. The larger the number of stator teeth and magnets, the smoother the output and the larger the torque produced. Of course this comes at the cost of larger size, greater mass and higher price. The large U8 BLDC motor

1 Fisher, P. (2014) *High Performance Brushless DC Motor Control*. BEng Thesis, May, CQ University, Australia.

from T-Motor has 36N 42P, a mass of 240 g, $K_v$ of 100, 135 or 170 and costs US$ 280.[2]

> ***Tip 7:*** To change the direction of travel of the motor, simply switch any two of the three output wires.

An example of what is inside a typical BLDC motor and its corresponding three-phase waveform is shown below and, in more detail, overleaf.

*Figure 4.1* The internal components of a typical multi-pole BLDC motor.[3]

The timing waveform shows how complicated the phase current supply is to each of the three phases. This makes it very difficult to measure accurately. In a sensorless BLDC motor, the back EMF is the feedback loop that provides the ESC with information about the position and speed of the rotor.[4] In this arrangement the control initially starts as open loop for the first pass, then remains in closed loop mode until stopped. With this design, slow speed control is somewhat problematic.

2 T-Motor (2018) *U-Type efficiency BLDC motors*. Available from: http://store-en.tmotor.com/category.php?id=38

3 DroneTrest (2018) *Brushless motors – how they work*. Available from: www.dronetrest.com/t/brushless-motors-how-they-work-and-what-the-numbers-mean/564

4 RCPlanes Online (2018) *Online Guide to Electrics*. Available from: https://rcplanes.online/guide5.htm

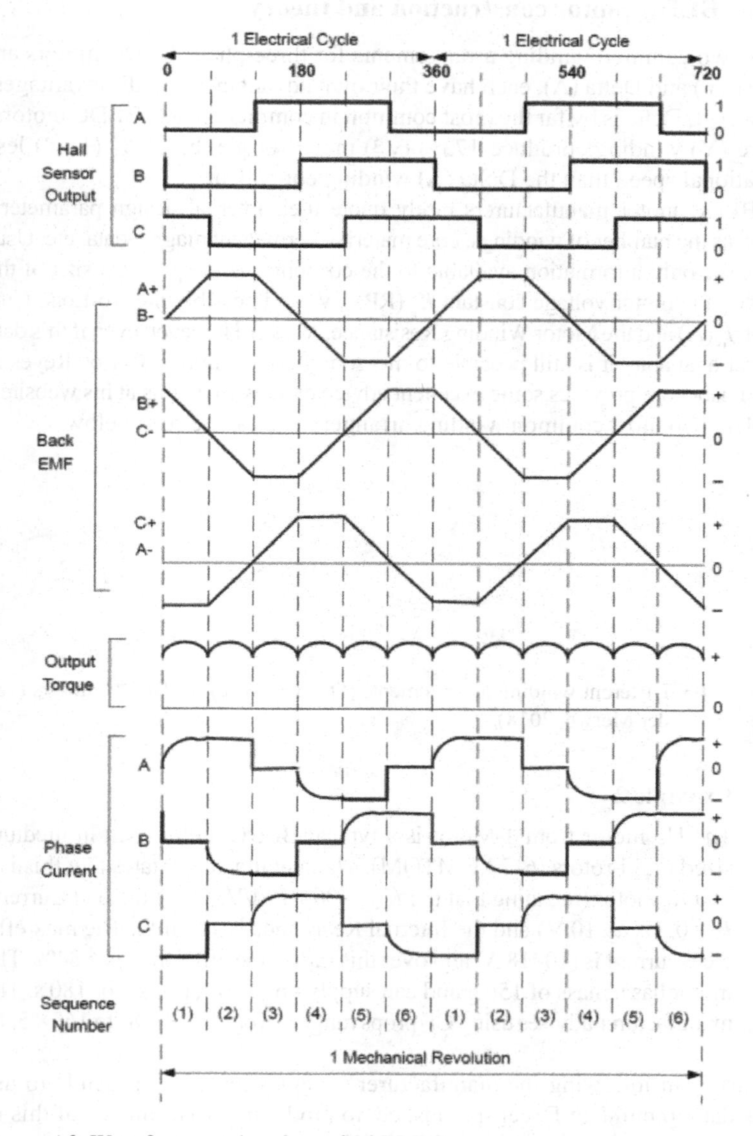


*Figure 4.2* Waveform graphs of a typical BLDC motor (©Microchip Technology Inc.).[5]

5 Yedamale, P. (2003) *AN885 Brushless DC (BLDC) Motor Fundamentals*. Microchip Technology Inc.

## 4.1 BLDC motor construction and theory

The two common winding arrangements for three-phase BLDC motors are Wye (Y) and Delta (Δ), each have their own advantages and disadvantages, however, Delta is by far the most common in commercial RC BLDC motors. Wye (Y) windings produce 173% ($\sqrt{3}$) more Torque, but 58% ($1/\sqrt{3}$) less rotational speed than the Delta (Δ) winding equivalent.

BLDC motor manufacturers rarely quote their internal design parameters, such as the number of windings, core material, permanent magnet data, etc. Usually, the only information available to the consumer is the physical size of the motor, the motor voltage constant $K_v$ (RPM/V) and possibly the No-Load Current, $I_0$ (A) and the Motor Winding Resistance, $R_m$ (Ω). However, even if this data is not available, it is still possible to measure these quantities. Carlos Reyes at rcadvisor.com provides some excellent advice on how to do this at his website.[6]

The two most common winding arrangements can be seen below.

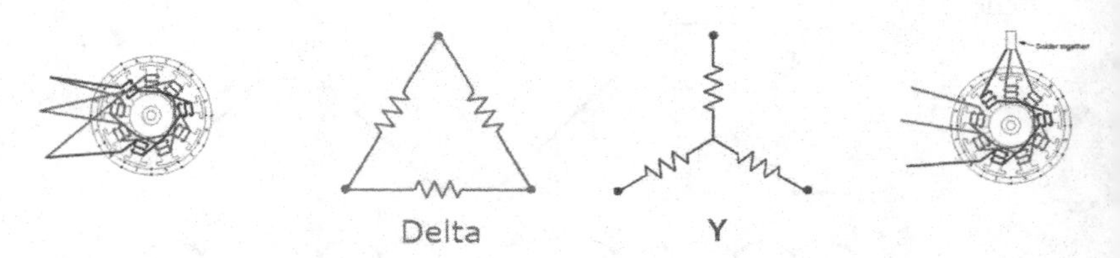


*Figure 4.3* Different winding arrangements (Delta vs Wye) for BLDC motors (van der Merwe, 2018).

**Example 2**

The U5 motor from T-Motor is a typical BLDC motor used in medium sized multi-rotors (6–7 kg MTOM). The manufacturer states that this is a 400 $K_v$ motor (meaning that the $K_v$ = 400 RPM/V), the No-Load Current, $I_0$ = 0.3 A @ 10 V) and the Internal Resistance is 116 mΩ. The max efficient current is (10–18 A) and over this range, the efficiency is > 84%. The motor has a mass of 156 g and can supply up to 30 A (max) for 180 s. The manufacturer advises using CF props ranging from 14″ × 4.8″ to 16″ × 5.4″.

Apart from following the manufacturer's suggestions, it is possible to use this data to build an Excel spreadsheet to model the performance of this or any other BLDC motor (see Appendix C).

6 Reyes, C. (2018) *Measuring motor constants* Available from http://rcadvisor.com/measuring-motor-constants-introduction.

***Tip 8:*** The three main BLDC motor constants ($K_v$, $I_0$ and $R_m$) are not actually true 'constants' and can vary with the manufacturing batch, applied voltage (see below) and temperature by about ±10% or more.[7]

Any model based on these values will therefore be a first-order approximation and the designer should not rely on the results as being an absolute representation of the performance of the BLDC motor.[8]

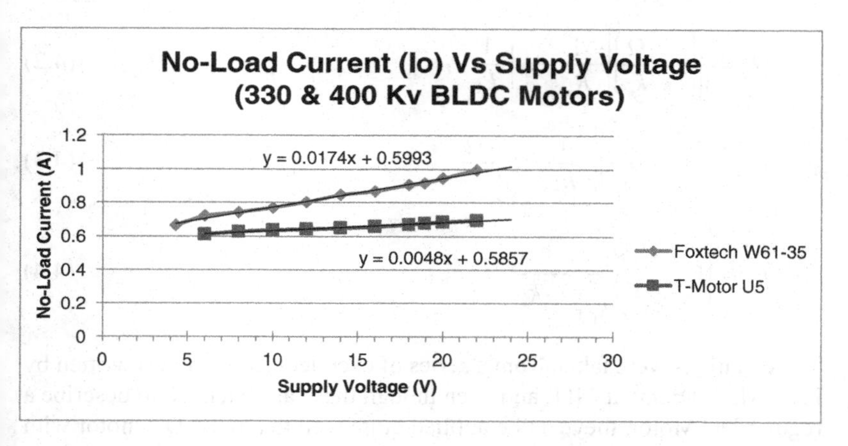


*Figure 4.4* A graph showing the variation of the no-load current, $I_0$ for various supply voltages.

T-Motor quoted an inaccurate no-load current value of 0.3 A @ 10 V for the U5 (400 $K_v$) BLDC motor, whereas Foxtech quoted an accurate no-load current value of 1.0 A @ 24 V for the W61-35 (330 kV) BLDC motor.

## 4.2 Equivalent brushed DC motor equations

The following equations were derived from DC Motor theory by Prof. Mark Drela at MIT and need to be handled with care when used to describe BLDC motors.

Li-Po Battery Power, $P_{elec} = v.i$
*Motor and Prop Speed,* $\Omega = (v - I.R_m).\ K_v$ *(where $K_v$ is in rad/s/Volt)*
*Motor Torque,* $Q_m = (i - I_0)/\ K_v$

7 Drela, M. (2006) *Second-Order DC Electric Motor Model.* Available from: http://web.mit.edu/drela/Public/web/qprop/motor2_theory.pdf

8 van der Merwe, C. (2018) *Motor Constants.* Available from: http://bavaria-direct.co.za/constants/

*Motor Shaft Power,* $Q_m \Omega = (i - I_0).(v - i.R_m)$
*BLDC Motor Efficiency,* $\eta_m$ = *Motor Shaft Power/Battery Power* = $(1 - I_0/i).(1 - i.R_m/v)$

*Taking these and resolving using v, and Ω, we can derive the following equations:*

$$i = \left[\left[\left(v - \frac{\Omega}{k_v}\right)\right] \cdot \frac{1}{R_m}\right] \tag{4.1}$$

$$Q_m = \left[\left[\left(v - \frac{\Omega}{k_v}\right)\right] \cdot \frac{1}{R_m} - i_o\right] \cdot \frac{1}{k_v} \tag{4.2}$$

$$P_{shaft} = \left[\left[\left(v - \frac{\Omega}{k_v}\right)\right] \cdot \frac{1}{R_m} - i_o\right] \cdot \frac{\Omega}{k_v} \tag{4.3}$$

$$\eta_m = \left[1 - \frac{i_o \cdot R_m}{v - \Omega/k_v}\right] \cdot \frac{\Omega}{v \cdot k_v} \tag{4.4}$$

The equations were taken from a series of excellent lecture notes written by Prof. Mark Drela[9] at MIT, and even though these are intended to describe a regular DC Motor, they can be applied quite well to any BLDC motor with some modifications and careful handling (see below).

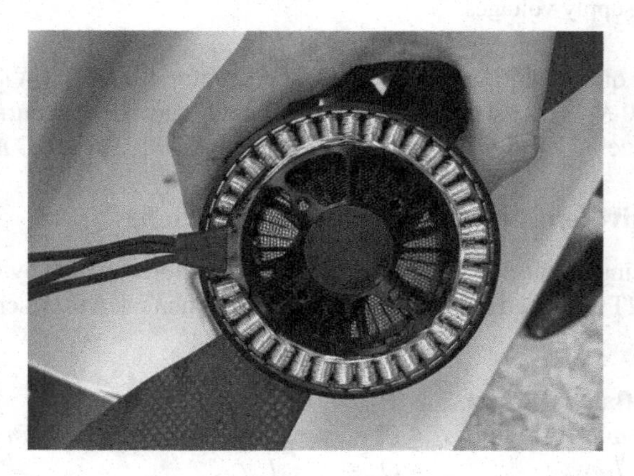

*Figure 4.5* Large multi-pole U8 BLDC motor from T-Motor.

9 Drela, M. (2005) *DC Motor/Propeller Characterization.* Lab 5 Lecture Notes, MIT, USA.

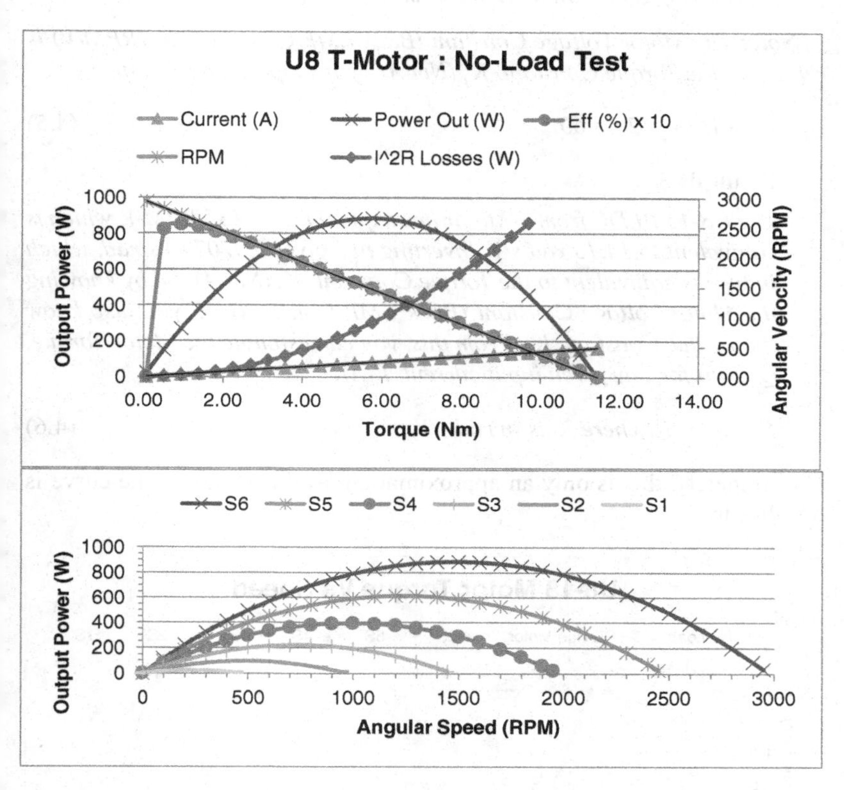


*Figure 4.6* Typical performance curves for a large (U8) BLDC motor.

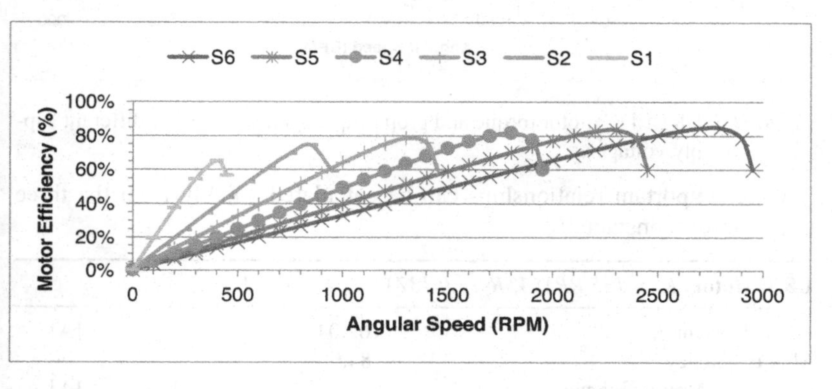


*Figure 4.7* Increasing max efficiency of the T-Motor U8 BLDC motor.

***Note:** The Motor Voltage Constant (Back EMF Constant), $K_v$ (RPM/V) is related to the Torque Constant, $K_T$ (Nm/A) by a simple relationship:*

$$K_T = 1/(K_v \cdot 2 \cdot \pi/60) \tag{4.5}$$

**Example 3**

*The U8-13 BLDC from T-Motor has a quoted $K_v$ = 135 RPM/V, which is equivalent to 14.13 rad/s/V. Inverting this gives 0.07074 Vs/rad, which in turn is equivalent to the Torque Constant, $K_T$ (Nm/A). So by knowing the Motor Voltage Constant (Back EMF Constant), $K_v$ you also know the Torque Constant, $K_T$. From this, you can estimate the Motor Torque, Q (Nm) for any given input current, I (A).*

$$K_T \cdot K_v = 1 \text{ (where } K_v \text{ is in rad/s/V)} \tag{4.6}$$

Unfortunately, this is only an approximation, as the actual torque curve is non-linear!

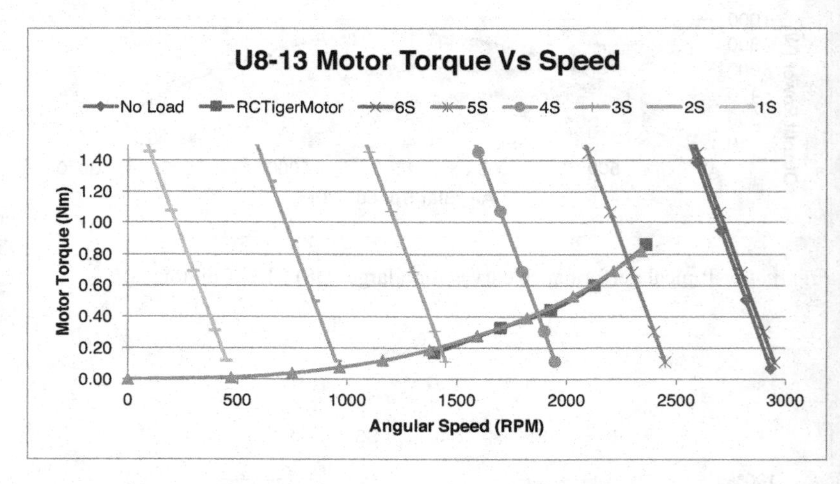


*Figure 4.8* U8 BLDC motor torque and prop torque against speed for different supply voltages.

Other important relationships can be calculated quickly from the three main motor 'constants':

| **U8 T-Motor:** $K_v$ = 135 RPM/V, $R_m$ = 0.137 Ω, $I_0$ = 1 A @ V = 22.2 V | | | |
|---|---|---|---|
| i-Stall Current | 162.04 | | [A] |
| Max Efficiency | 84.9% | | |
| Current @ Max Efficiency | | 12.73 | [A] |
| Angular Velocity @ Max Efficiency | | 2761 | [RPM] |

The previous useful parameters were derived using the equations listed below:

$$\text{i-Stall Current} = V/\, R_m = 162.04\ A \tag{4.7}$$

$$\text{Maximum Efficiency} = \left[1 - \sqrt{\left(\frac{I_0}{i - Stall}\right)}\right]^2 = 84.9\ \% \tag{4.8}$$

$$\text{Current (A) @ Maximum Efficiency} = \sqrt{(I_0 . i - Stall)} = 12.73\ \text{A} \tag{4.9}$$

$$\text{Angular Velocity (RPM) @ Max Efficiency} = \frac{No\,Load\,RPM}{1 + \sqrt{\dfrac{I_0}{i - Stall}}}$$

$$= \frac{2932}{1 + \sqrt{\dfrac{1}{162.04}}} = 2761\ \text{RPM} \tag{4.10}$$

$$K_M = {K_T}\Big/{\sqrt{R_M}} = {0.07074}\Big/{\sqrt{0.137}} = 0.1911\ \text{Nm/W} \tag{4.11}$$

If you know the design parameters of the motor, you can use the traditional motor design equations:

$$Back\ EMF = N.l.r.B.\omega \tag{4.12}$$

Where:
*N = Number of winding turns per phase. L = Length of the motor (m). r = Internal radius of the rotor (m). B = Rotor magnetic field density (Tesla). ω = Angular velocity of the motor (rad/s).*

A good overview of BLDC motor theory is provided by David Wall in his MSc thesis and in a book chapter by Zhu and Watterson.[10,11]

$$\omega_r = \frac{V_{motor}}{p\dfrac{\lambda_m}{2}} - \frac{R_1}{m\left(p\dfrac{\lambda_m}{2}\right)^2} T_{em} \tag{4.13}$$

$$T_{em} = \frac{mp}{2}\lambda_m I \tag{4.14}$$

10 Wall, D.L. (2012) *Optimum Propeller Design for Electric UAVs*. MSc. Thesis, Auburn University, Alabama, USA. Available from: https://etd.auburn.edu/bitstream/handle/10415/3158/David_Wall_Thesis.pdf?sequence=2

11 Zhu, J.G. & Watterson, P., *Electromechanical Systems – Chapter 12. Brushless DC Motors*. Lecture Notes. Available from: http://services.eng.uts.edu.au/cempe/subjects_JGZ/ems/ems_ch12_nt.pdf

Where:

$V_{motor}$ = *Motor voltage (V). P = Number of pole pairs. m = number of phases. I = Motor current (A).*

$T_{em}$ = *Electro-magnetic torque (Nm).* $R_i$ = *Motor internal resistance (ohm).*

$\lambda_m$ = *Flux Linkage of the stator windings (Wb).*

## 4.3 Comparing theoretical and empirical data

In the previous technical analysis of theoretical brushed and brushless dc motor equations, it is important to stress the role of the % throttle. Here, the only correct value will be obtained at 100% throttle, i.e. where the ESC is effectively providing the whole supply voltage to the windings of the motor. Since this is a PWM controlled system, 100% throttle relates to fully on and 0% throttle to fully off. Therefore, the 50% throttle mark is equivalent to exactly half the supply voltage.

For a 6S system (22.2 V) operating at 50% throttle, the effective supply voltage is actually 11.1 V. Since the % throttle is approximated to be linear with motor speed.

A further point to note is that although the motor torque constant, $K_T$ (Nm/A) can be calculated directly from the given motor voltage constant, $K_v$ (RPM/V) (see Equation 4.6), it is not a constant and changes with motor speed and supply voltage (see below).

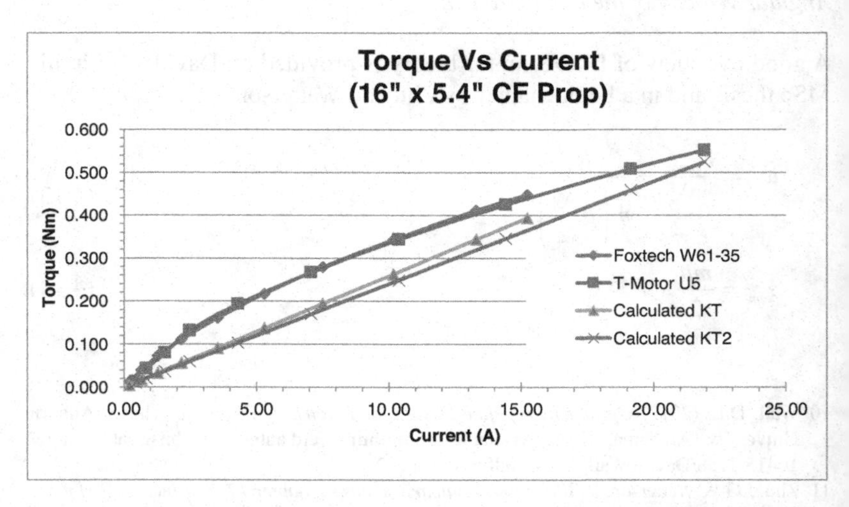


*Figure 4.9* Comparison of BLDC motor torque with current showing the non-linear relationship.

Having analysed and compared the theoretical equations with actual motor/prop empirical data, several useful observations can be made:

- An accurate value of $K_v$ is essential and has a large effect on several equations.
- We cannot rely on the motor torque constant, $K_T$ which is derived directly from $K_v$, as this is dependent on the supply voltage and is non-linear as shown earlier.
- The motor winding resistance, $R_m$ is not a constant and varies with temperature.
- The no-load current, $I_0$ is not a constant and varies with supply voltage, which in turn is dependent on the BLDC motor specification (bearing friction, ventilation, hysteresis, etc.).
- Using the % throttle to vary the supply voltage gives good agreement (< 15%) for the angular speed equation.
- Using a tuned % throttle approach, can reduce the angular speed error to < 1%.
- The % throttle data points are very sensitive and even a small increase or decrease (1–2%) can have a big effect on the result.
- Most data points in the range 10–40% throttle are inaccurate and should not be relied upon.

### 4.3.1 *Analysis of motor torque equations*

Five different motor torque equations were developed to try to predict the motor torque (Nm), given other motor constants:

---

Torque Calc 1 = $K_T \cdot$ I (very inaccurate (Average of 25% under prediction for 50–100% throttle))

Torque Calc 2 = Shaft Power (Calc 1)/Measured Angular Velocity (rad/s)

*(Note: Shaft Power (Calc 1)* $= (V - (I \cdot R_m)) \cdot (I - I_0)$*)*

Torque Calc 3 = Shaft Power (Calc 2)/Measured Angular Velocity (rad/s)

*(Note: Shaft Power (Calc 2) = Electrical Power Used –* $((I\char`^2 \cdot R_m)+(V \cdot I_0))$*)*

Torque Calc 4 = Shaft Power (Calc 3)/Measured Angular Velocity (rad/s)

*(Note: Shaft Power (Calc 3)* $= (I - I_0 - (V/R2)) \cdot (V - (I \cdot R_m))$*)* (NB: Almost the same result as Calc 2)

(Taken from the four constant model)[12]

Torque Calc 5 = Uses trend lines from plots of measured torque values (50–100% throttle). (Most accurate method)

---

12 Du Plessis, F. (2011) *Brushless DC Motor Characterisation and Selection for a Fixed Wing UAV*. Available from: www.academia.edu/1360191/Brushless_DC_Motor_Characterisation_and_Selection_for_a_Fixed_Wing_UAV

*Table 4.1* Calculated motor torque error using five different methods.

| | *Torque (Nm) Error* | | | | | | | |
|---|---|---|---|---|---|---|---|---|
| | *Calc 1* | *Calc 2* | *Calc 3* | *Calc 4* | *Calc 5* | $K_v$ | $I_0$ | $R_m$ |
| **W61-35** | *−19.2%* | *−0.7%* | *−3.5%* | *−2.8%* | *0.38%* | ***330*** | ***0.997*** | ***0.095*** |
| **U5** | *−16.7%* | *7.0%* | *6.5%* | *6.5%* | *0.42%* | ***370*** | ***0.695*** | ***0.116*** |
| | *NB: % Throttle 50–100%* <br> *NB: Average % error over throttle range* | | | | | *NB: Changing these constants will change the error table.* | | |
| | The W61-35 appears to perform best when modelled as a 330 $K_v$ motor. <br> The U5 appears to perform best when modelled as a 370 $K_v$ motor. | | | | | | | |

Using the previous equations, an analysis of two BLDC motors (Foxtech W61-35 & T-Motor U5) when combined with a T-Motor 16″ × 5.4″ CF propeller was conducted. The results can be observed in Table 4.1.

As can be seen in Table 4.1, applying the Torque Calc 5 method provided the most accurate values for the motor torque. Perhaps this is no surprise, given that these were derived from motor test trend lines. Apart from method 1, the other three methods performed reasonably well, with some variation from motor to motor.

All methods, with the exception of method 1, require some *a-priori* test data. So it is up to the reader how much effort they are willing to apply and what level of accuracy they can accept.

### 4.3.2 *Analysis of motor speed equations*

Three different motor speed equations were developed to try to predict the motor speed (RPM), given the other motor 'constants':

| | |
|---|---|
| Calc Speed 1 = $(((V/100 \cdot \%\text{Throttle})/K_T) - ((R_m \cdot Q_m)/((K_T\text{^}2))))/(2 \cdot \text{PI}()/60)$ | *NB: Torque ($Q_m$) is required for this equation.* |
| Calc Speed 2 = $(((V/100 \cdot \%\text{Throttle}) - (I \cdot R_m))/(K_T))/(2 \cdot \text{PI}()/60)$ | |
| | *NB: Calc Speeds 2 & 3 results are effectively identical.* |
| Calc Speed 3 = $K_v \cdot ((V/100 \cdot \%\text{Throttle}) - (I \cdot R_m))$ | |

*Table 4.2* Calculated motor speed error using three different methods.

| | *Angular Velocity (RPM)* | | | | | |
|---|---|---|---|---|---|---|
| | *Speed1* | *Speed2* | *Speed3* | $K_v$ | $I_0$ | $R_m$ |
| **W61-35** | *0.6%* | *1.7%* | *1.7%* | ***330*** | ***0.997*** | ***0.095*** |
| **U5** | *−1.0%* | *0.6%* | *0.6%* | ***370*** | ***0.695*** | ***0.116*** |
| | *NB: % Throttle 50–100%* *NB: Average % error over throttle range.* | | | *NB: Changing these constants will change the error table.* | | |
| | The W61-35 appears to perform best when modelled as a 330 $K_v$ motor. The U5 appears to perform best when modelled as a 370 $K_v$ motor. | | | | | |

Using the previous equations, an analysis of two BLDC motors (Foxtech W61-35 & T-Motor U5) when combined with a T-Motor 16″ × 5.4″ CF propeller was conducted. The results can be observed in Table 4.2.

As can be seen in Table 4.2, all three methods provided very accurate values for the motor speed under load conditions. Method 1 was the most accurate, but required the motor torque value. Methods 2 and 3 are equal, with method 3 being computationally simpler.

### 4.3.3 *Test measurements*

In this test, the torque was being measured using an RTS-100 sensor with a calibrated display. The Current was measured using an oscilloscope with a Fluke current probe at the ESC input. The motor speed (RPM) was measured using a handheld laser RPM meter. The supply voltage was measured in two places: (1) ESC supply (Fluke Multimeter), and (2) ESC output, one of the motor phases (oscilloscope).

The same ESC was used for both motor tests (MayTech 40A-Opto). Voltage varied with bench power supply from 4.3 V (ESC cut off) to 19 V (Power supply top end). The motor winding resistance, $R_m$ was measured using a constant power 1 A and 1 V. The voltage drop is measured across all three phases, then averaged out and calculated back to resistance using Ohms law.

From the manufacturer's website, these two BLDC motors were quoted as having the following motor 'constants'.

*Table 4.3* Manufacturer quoted motor 'constants'.

| *BLDC Motor* | *Manf. $K_v$* | *Measured $K_v$* | *Tuned $K_v$* | *No-Load Current, $I_0$* | *Winding Resistance, $R_m$* |
|---|---|---|---|---|---|
| Foxtech W61-35 | 330 | 370 | 330 | 1.0 A @ 24 V | 0.052 mΩ |
| T-Motor U5 | 400 | 398 | 370 | 0.3 A @ 10 V | 0.116 mΩ |

As stated previously, the manufacturer's $K_v$ value can vary by ±10% or more. Therefore, the data is perhaps not a surprise. However, if the designer used the quoted motor 'constant' values in the previous equations, the error values would over-predict motor torque by 11% and motor speed by 9%.

### 4.3.4 *Online design tools*

These results are still regarded as being 'good enough' for most purposes. Over the years, several software programs, both freeware and licenced paid have been developed for use by RC users based on the previous equations.[13,14,15] There is now also a design package specifically for multi-rotor enthusiasts.[16]

Most of these programs have a mixed freeware element, with more components available within the licenced paid version. As can be seen below, the xcopter eCalc version quotes an accuracy of ±15%.

The user selects the basic configuration and then the individual propulsion elements from a series of drop-down menus. Of course, the component lists are not exhaustive, with some only available from within the licenced paid version. Another issue with these design tools is the reliance on manufacturer's data to give accurate predictions. As I have stated many times already, this data can sometimes be inaccurate, missing and even misleading. However, the user does have the option of manually entering their own data based on measurements, if the component is not available to select.

13 MotorCalc (2018) *Online calculator*. Available from: www.motocalc.com/
14 eCalc (2018) *Multi-rotor online software calculator*. Available from: www.ecalc.ch/
15 DriveCalc (2018) *Online calculator*. Available from: www.drivecalc.de/
16 eCalc (2018) *Multi-rotor online software calculator*. Available from: www.ecalc.ch/xcoptercalc.php

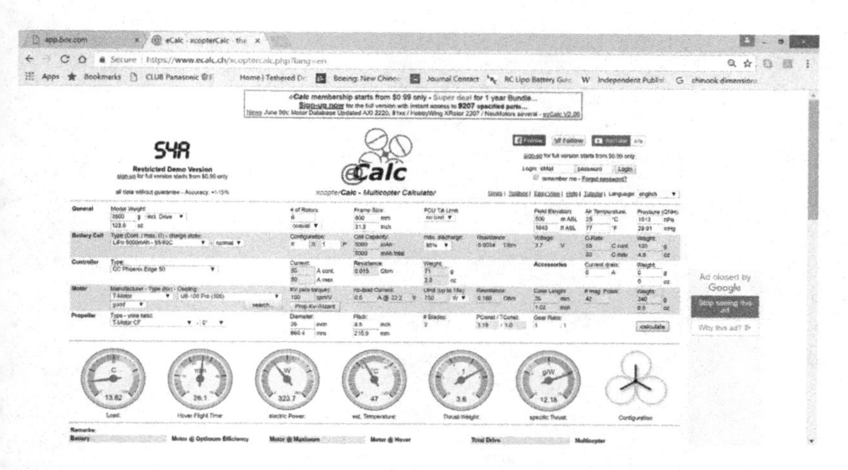


*Figure 4.10* The main screen of the xcopterCalc multi-rotor design package (eCalc, 2018).

The good thing about these software packages is that they are available online for a relatively low cost or even free. The bad thing is that few people truly understand what equations stand behind the software. Another issue is that the motor/prop combination required is sometimes the one which is not free, or which has not been added to the database due to the fact that it is so new. Again, you can enter data manually, if known. This is then used to compute the outputs, which is a helpful tool.

Ultimately, any guide or tool which provides a ball park figure of performance is useful before one embarks on the purchase of expensive hardware components from a distant supplier.

# 5 The propeller (rotor)

A propeller or Rotor in a rotary-wing aircraft is equivalent to the wing of a fixed-wing aircraft. What appears to be the simplest of the four main propulsion components is actually one of the most critical, complex and misunderstood.[1]

The history of aviation, since the Wright brothers first took to the air in 1903, has seen many developments. In fact, most of the theoretical understanding of the new science of aerodynamics was developed in the first half of the 20th century. The Wright brothers had by 1905 developed a so-called 'bend end' 8.5 feet diameter propeller with an efficiency of 82%. Even today, the world's best full-scale propeller manufacturers struggle to reach 90% efficiency. With so much early success, this area of science has almost become a forgotten art.

The propeller (rotor) converts the rotational output power of the BLDC motor into the movement of air using the principle of Newton's 3rd Law of Motion – *for every action there is an equal and opposite reaction.* Basically, the energy of rotation is converted into the acceleration of air below the disk plane of the propeller (rotor), this Thrust (N) is the force that lifts the multi-rotor aircraft, and pulls the fixed-wing aircraft forwards. The Simple Momentum Theory (SMT) equations presented in Section 5.4, dictate in a simplistic way, how well this can be achieved.

Scale props used in small unmanned aircraft are typically fairly inefficient, ranging from approx. 30–70%, with the efficiency being directly proportional to the diameter of the prop (see below). Props used in small multi-rotors range from 2″ to 40″ in diameter, with 9–18″ being the most common.

1 Curtiss-Wright (1944) *Propeller Theory.* Available from: https://whirlwindpropellers.com/aircraft/propeller-theory-1944/

**Example 4**

A BLDC motor operating at an efficiency, $\eta_m = 85\%$ (very good) when combined with a propeller operating with an efficiency, $\eta_p = 80\%$ (excellent), will have an overall efficiency of:

$$\eta_{overall} = \eta_m \times \eta_p = 0.85 \times 0.8 = 0.68\ (68\%) \tag{5.1}$$

Even in this 'best' scenario (which usually only occurs over a very narrow operational speed range), 32% of the total input energy is lost within the propulsion system, mostly within the propeller.

One of the most reliable sources of small propeller data is the database from the University of Illinois at Urbana (UIUC), developed by researchers under Professor Michael Selig.[2] The database currently holds results for about 210 props (Vol. 1: 140 props & Vol. 2: 70 props) ranging in diameter from 1.85″ to 19″, with the vast majority in the range 7–11″ diameter.

The figure below clearly shows that even props with the same diameter and pitch can have varying efficiencies, ranging from 42% to 63% for the 11″ propellers.

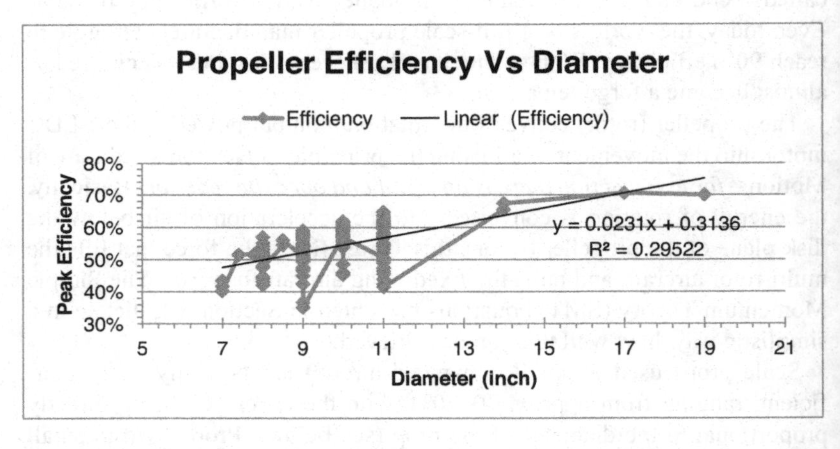


*Figure 5.1* Graph showing increasing propeller peak efficiency with increasing propeller diameter.

2 Brandt, J.B., Deters, R.W., Ananda, G.K. & Selig, M.S. (1 January 2016) *UIUC Propeller Database*. University of Illinois at Urbana-Champaign. Available from: http://m-selig.ae.illinois.edu/props/propDB.html

## 5.1 Propeller or rotor?

The question of how to refer to this vital aerodynamic component might seem esoteric, but is quite interesting and important. A propeller is generally associated with a fixed-wing aircraft, whereas a rotor is generally associated with a rotary-wing aircraft (sometimes referred to as a rotorcraft). A multi-rotor has more in common with a helicopter, which is of course a member of the rotorcraft family; this type of aircraft, almost universally uses variable pitch rotors, rather than fixed pitch propellers which are used in the vast majority of multi-rotors. The use of propellers in a multi-rotor is therefore a bastardisation of their intended role. However, they do seem to work!

*Figure 5.2* The Boeing CH-47 Chinook helicopter showing its airfoil rotor blade design.

Propellers were originally designed and intended for fast forward flight at constant cruise speed in fixed-wing aircraft and their use, in essentially, a hovering VTOL aircraft is counter-intuitive, nevertheless, they appear to work quite well. However, understanding their efficiency is the key to optimising their performance.

## 5.2 Propeller variables

Propellers come in all shapes and sizes, are handed, either right or left-handed. They range from single-bladed to four or even five-bladed (two-bladed being the most common), they are usually fixed pitch, but can be variable pitch, such as in helicopter rotors. They can also be foldable. The latest research is looking into the design of variable geometry, variable material props, perhaps with embedded sensors or actuators. Bird and insect wings (Ornithopters) are an obvious inspiration to designers of the next generation of propellers.

Manufactured from Wood, Plastic or Carbon Fibre, the cost of scale propellers can range from £ 2–300 each depending on their size and materials used. The latest designs tend to be manufactured from Carbon Fibre which gives them a light, but very strong design. The gap between scale props and full-sized propellers (> 60″ diameter) is becoming increasingly narrow as larger propellers (up to 40″ diameter as of June 2018) become increasingly available to the RC enthusiast.[3]

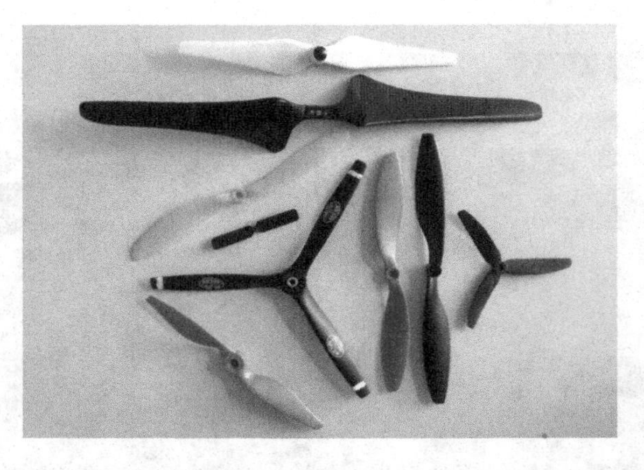

*Figure 5.3* Various propeller designs showing the range of materials and geometry.

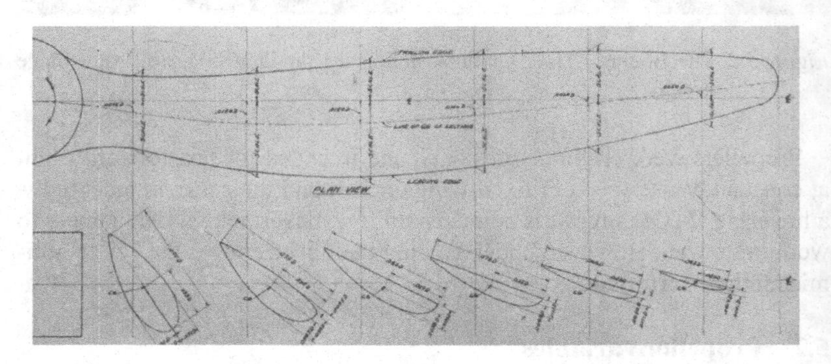

*Figure 5.4* Propeller sections at r/R in the spanwise direction.

3 Xoar International (2018) *Multicopter Propellers*. Available from: www.xoarintl.com/multicopter-propellers/

At this point it is probably a good idea to state that a fixed pitch propeller is essentially a series of different pitches joined together to give a blended profile, as can be seen from the previous figure.[4]

Propellers are usually denoted in inches and designated as the diameter multiplied by the pitch. Manufacturers quote prop sizes such as 5″ × 3″, 16″ × 5.4″ and 40″ × 13.1″. Props are now being specifically designed for use in multi-rotor aircraft, and as such they have large diameters for efficiency and Pitch/Diameter (*P*/*D*) ratios in the region of 0.3–0.35 for optimum hovering efficiency and slow forward flight velocity. This can be compared to a prop designed for efficient fast forward cruise speed, such as the APC 11″ × 10″ (*P*/*D* ratio = 0.91) and peak efficiency of 76% at J = 0.67 (see below).

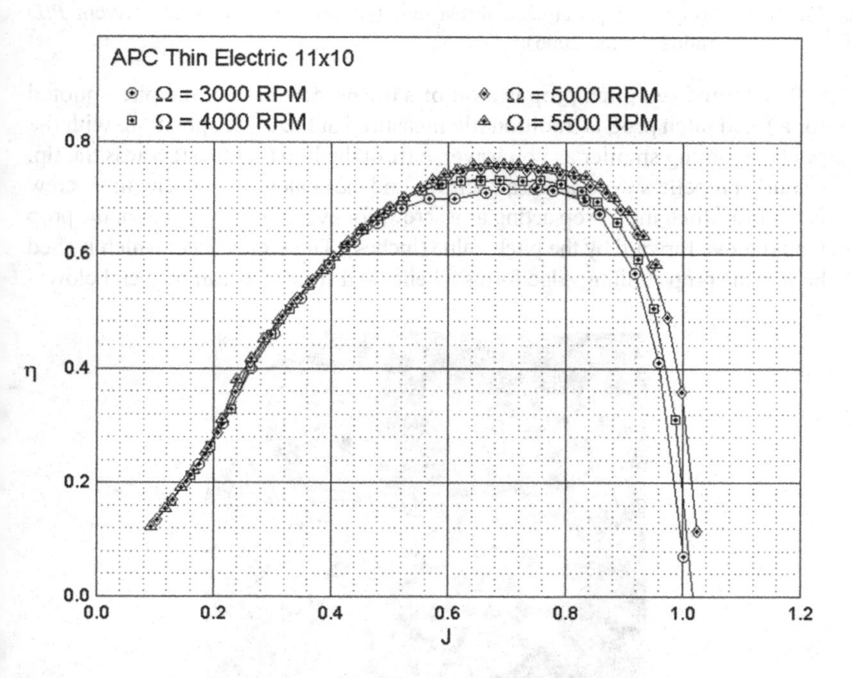


*Figure 5.5* Typical propeller data showing the relationship between efficiency and advanced ratio, J.[5]

4 Weick, F.E. (1925) *Simplified Propeller Design for Low-Powered Airplanes*. NACA Technical Note 212.

5 Brandt, J.B., Deters, R.W., Ananda, G.K. & Selig, M.S. (June 2018) *UIUC Propeller Database*. University of Illinois at Urbana-Champaign. Available from: http://m-selig.ae.illinois.edu/props/propDB.html

The figure below shows Thrust, $T$ (N) vs Velocity, $V$ (m/s) for two different cases (low and high $P/D$ ratios). As stated, multi-rotor propeller designs tend to have $P/D$ ratios in the range of 0.3–0.35.

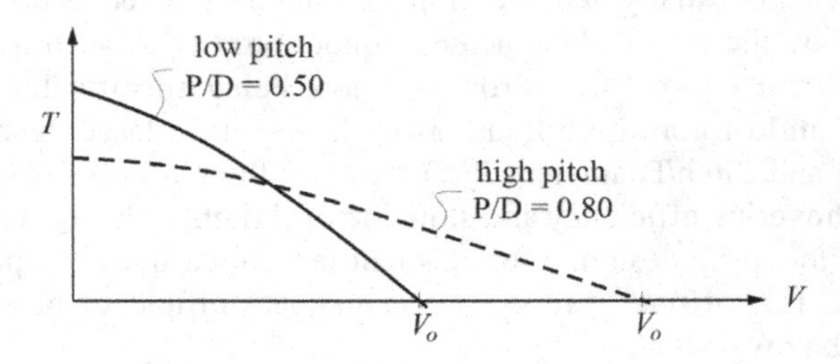


*Figure 5.6* Relationship between thrust and forward velocity with different $P/D$ ratios (Drela, 2006).[6]

The term fixed pitch prop is a bit of a misnomer; in fact, the pitch quoted for a fixed pitch prop is traditionally measured at the 0.75R position, with the pitch becoming smaller as you traverse from the hub (centre) towards the tip.

Early propellers were called air screws, and the analogy to a mechanical screw is not lost.[7] Imagine a prop acting as a screw, for every 360° of rotation, the prop would move forward by the pitch value (inches). The vortex sheet which is shed behind the props trailing edge forms a helicoidal surface as can be seen below.

*Figure 5.7* Vortex sheet helicoidal surface representation shed from the propeller's trailing edge.

6 Drela (2006) *DC Motor/Propeller Characterization.* Available from: https://ocw.mit.edu/courses/aeronautics-and-astronautics/16-01-unified-engineering-i-ii-iii-iv-fall-2005-spring-2006/systems-labs-06/spl3.pdf

7 Larrabee, E.E. (1980) The screw propeller. *Scientific American*, 243(1), 134–149.

## 5.3 Propeller geometry

Given the pitch value at the 0.75R location, the whole range of spanwise pitches can be generated using a simple mathematical formula in Excel:

$$\text{Twist Angle (deg)} = \text{DEGREES(ATAN(Pitch/(2} \cdot \text{PI()} \cdot \text{r)))} \quad (5.2)$$

***Tip 9:*** If you use (mm) for the radius (r), then use (mm) for the pitch (5.4″ = 137.16 mm).

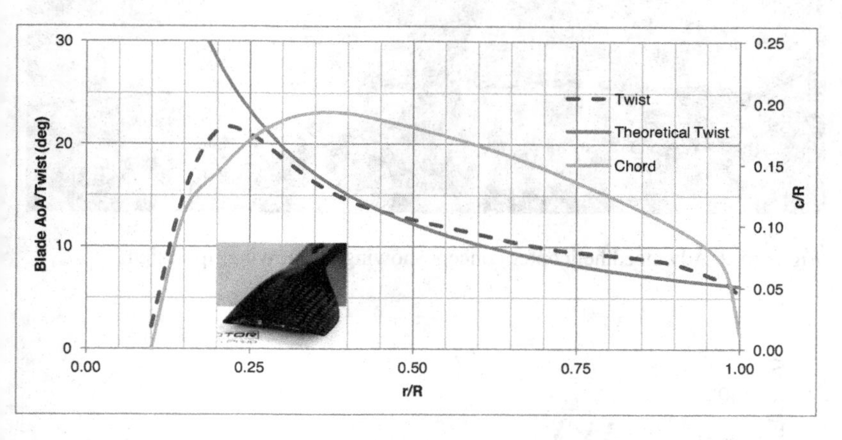


*Figure 5.8* Blade geometry for a T-Motor 16″ × 5.4″ carbon fibre propeller.

Fixed pitch props vary in pitch and chord spanwise; they twist and taper as well as sometimes rise or fall at the tip to form a small winglet.

Variable pitch rotors on the other hand generally have constant pitch and chord spanwise, instead relying on the variable pitch mechanism to adjust the pitch (AoA) of the rotor (via collective or cyclic input). They can also articulate up and down (flap), as well as move side to side (lead/lag sometimes called feathering).

The latest designs of full-sized rotor often have an advanced tip design to reduce tip losses and noise (see below). This was originally developed in the UK by BERP in the mid-1980s and was used to gain the world speed record 216 knot (111 m/s) for the Lynx helicopter on 11 August 1986.

Multi-rotor designs generally favour the simplicity of the fixed pitch propeller over the complexity of the variable pitch rotor. However, for true aerodynamic performance, including inverted flight, the variable pitch rotor design is essential.

*Figure 5.9* Advanced helicopter concept showing swept wing tip design.

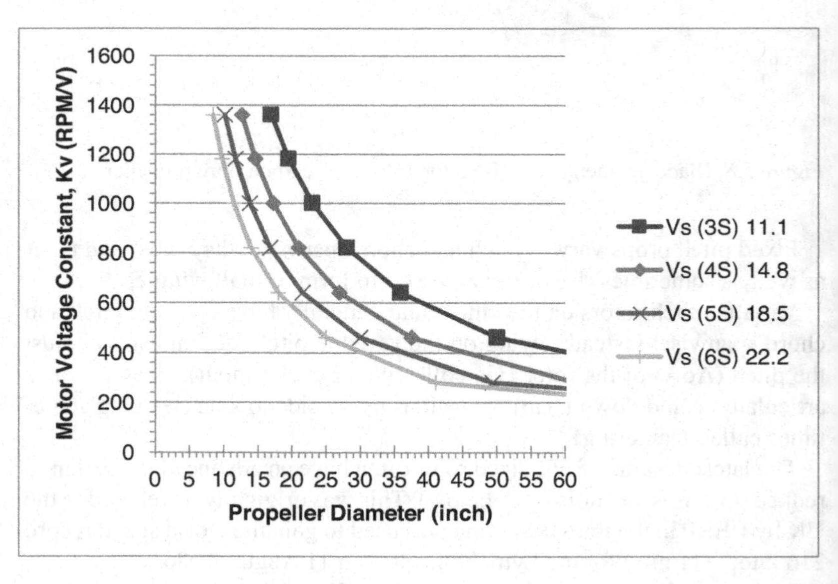


*Figure 5.10* Graph showing the limits of propeller diameter for different supply voltages and $K_v$ values.

## 5.4 Simple momentum theory

For a basic understanding of the action of a propeller (rotor) on a fluid (air), we can make use of the Simple Momentum Theory (SMT), sometimes referred to as the Actuator Disk Theory. Originally developed by Rankine, Froude and others in the late 19th century to describe marine propellers, this theory is only concerned with the balance of mass, momentum and energy. It does not try to analyse the flow around the surface of the individual blades, and therefore is a gross approximation.

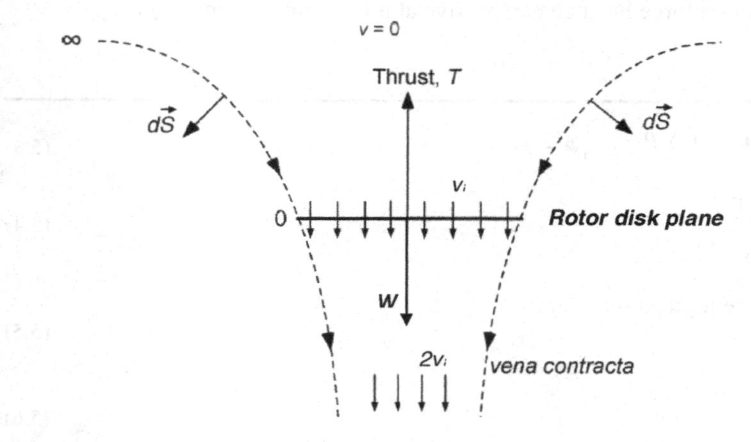


*Figure 5.11* 2D representation of the rotor flow field using SMT (© Cambridge University Press).[8]

Assumptions:

- The rotor is modelled as an actuator disk which adds momentum and energy to the air.
- There is no inflow or outflow through the side boundary, dS.
- The flow is incompressible.
- The flow is steady, inviscid and irrotational.
- The flow is one-dimensional and uniform.
- There is no swirl in the wake.

8 Leishman, J.G. (2016) *Principles of Helicopter Aerodynamics*, 2nd edition. Cambridge University Press, Cambridge.

Flow takes the form of the classic hour glass phenomena, as shown in the previous figure. The induced velocity of the air across the rotor disk plane is $v_i$. In the hover condition, at a point far above the rotor the air velocity, $v$ is zero.

Through the use of the conservation of momentum equations, the mass flow rate, $\dot{m}$ at any point in the system must be equal and therefore it can be shown that the far wake velocity is $2v_i$. The far wake area (vena contracta) is therefore half the rotor disk plane area (A/2). In reality this is slightly higher than 0.5A, and empirically has been calculated as closer to 0.7A.

From a force balance perspective at hover (substituting $v_i$ *for* $v_h$:

$$T = W = A.\Delta P = A\frac{1}{2}\rho(2v_h)^2 \tag{5.3}$$

$$v_h = \sqrt{\frac{T}{2A\rho}} \tag{5.4}$$

Considering a power balance:

$$P = Tv_h \tag{5.5}$$

Therefore:

$$P = Tv_h = 2\rho A v_h^3 \tag{5.6}$$

Another quantity of interest is the induced inflow ratio, $\lambda_i$ *or* $\lambda_h$

$$v_h = \lambda_h \Omega R = \lambda_h v_{tip} \tag{5.7}$$

The Disk Loading (DL) and Power Loading (PL = *T*/*P* (0.03–0.06 N/W) are also of great interest:

$$DL = \frac{T}{A} = Typically(240 - 480)N/m^2 \tag{5.8}$$

In the following section, a range of non-dimensional coefficients will be introduced, which will enable the rotor (propeller) designer to compare one design against another in a fair and objective way.

Note: There is equivalence between the rotor and propeller equations. It is also important to point out that the (½) in the denominator differentiates rotors used in the UK, Europe and Russia from those in the US. Therefore, the values of $C_T$, $C_P$ and $C_Q$ will be a factor of two greater than their US equivalent values.

## 5.5 Non-dimensional rotor (propeller) coefficients

$$C_T \text{ Rotor Thrust Coefficient} = \frac{T}{\frac{1}{2}\rho A(\Omega R)^2} \tag{5.9}$$

$$C_{Tp} \text{ Propeller Thrust Coefficient} = 3.875C_T = \frac{T}{\rho n^2 D^4} \tag{5.10}$$

$$C_P \text{ Rotor Power Coefficient} = \frac{P}{\frac{1}{2}\rho A(\Omega R)^3} \tag{5.11}$$

$$C_{pp} \text{ Propeller Power Coefficient} = 12.175\, C_p = \frac{P}{\rho n^3 D^5} \tag{5.12}$$

$$C_Q \text{ Rotor Torque Coefficient} = \frac{Q}{\frac{1}{2}\rho A(\Omega R)^2 R} \tag{5.13}$$

$$C_{Qp} \text{ Propeller Torque Coefficient} = 1.938\, C_Q = \frac{Q}{\rho n^2 D^5} \tag{5.14}$$

It is important to note here the quantities used in calculating these non-dimensional coefficients:

| | |
|---|---|
| *A Rotor disk area ($m^2$)* | *Q Rotor (Propeller) torque (Nm)* |
| *c Blade chord (m)* | *R Radius of rotor blade (m)* |
| *D Propeller diameter (m)* | *T Rotor (Propeller) thrust (N)* |
| *n Propeller rotational speed (rev per second)* | *$\rho$ Air density (1.225 kg/$m^3$ at sea level)* |
| *$N_b$ Number of blades* | *$\sigma$ Rotor solidity (for one rotor), $N_b c/\pi R$* |
| *P Rotor (Propeller) power (W)* | *$\Omega$ Rotational speed of the rotor (rad/s)* |

These coefficients are used by the engineer to assess the performance of the system and ultimately the efficiency of the propulsion system.

***Tip 10:*** Note that the Thrust is proportional to the square of the Angular Velocity, and the Output Power is proportional to the cube of the Angular Velocity (as seen below).

The efficiency of a propulsion system can be calculated in two different ways depending on whether this is a rotor or a propeller driven system. Basically, rotary-wing aircraft, such as a helicopter or multi-rotor, that spend most of their time hovering, use an efficiency metric called the Figure of

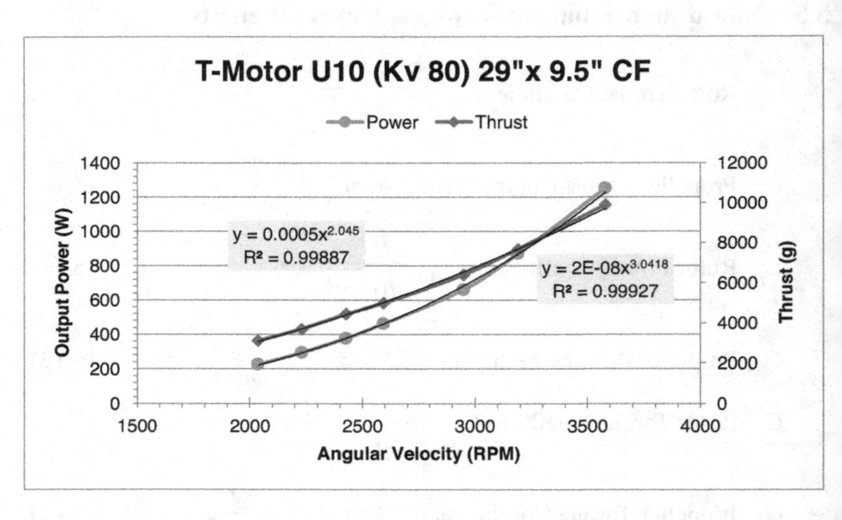


*Figure 5.12* Graph showing the $n^2$ relationship of thrust and $n^3$ relationship to output power.

Merit (FM). Whereas, a fixed-wing aircraft (propeller driven) would use a propeller efficiency term, $\eta_p$.

## 5.6 Figure of Merit (FM)

The Figure of Merit (FM) was introduced in the first half of the 20th century by Bruce (1903) and developed further by Glauert (1935) and Prewitt (1941), who was working for the Kellett Aircraft Company. This is a ratio of the ideal power (hover) to the actual power required to hover:

$$\text{Figure of Merit (FM) (single rotor)} = \frac{ideal\ power}{actual\ power} = \frac{Tv_i}{P} = \frac{C_T^{3/2}}{\sqrt{2}\,C_p} \quad (5.15)$$

A very efficient helicopter might have an FM = 0.8. However, hover efficiency and fast forward flight capability are always in conflict with each other, therefore an FM = 0.6 might infer a greater top speed at the expense of hover efficiency.

It has been found using the Blade Element Theory, that the FM is a function of $\frac{C_l^{3/2}}{C_d}$ of the airfoil, maximising this ratio will also maximise the FM.

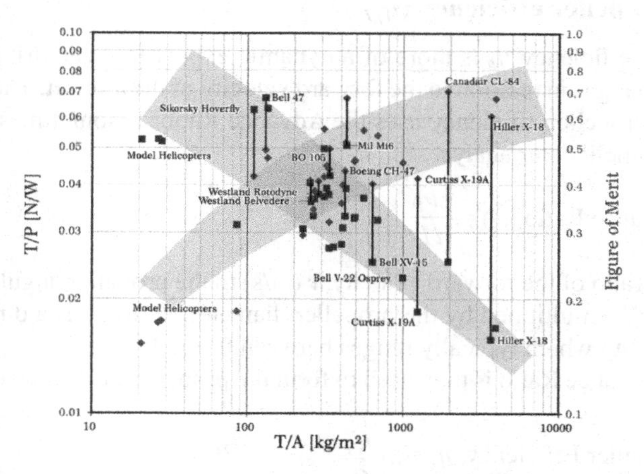


*Figure 5.13* Figure of Merit of various VTOL aircraft mapped against thrust and power loading.[9]

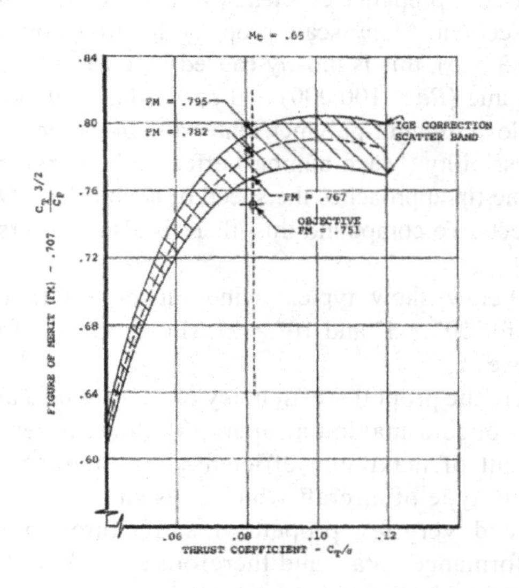


*Figure 5.14* Figure of Merit (FM) for an early design of Boeing CH-47 Chinook rotor blade.[10]

9 Hepperle, M. (2018) *Helicopters and Compound Aircraft*. Available from: www.mh-aerotools.de/airfoils/helico1.htm

10 Boeing Vertol Company (1977) *Heavy Lift Helicopter – Advanced Technology Component Program – Rotor Blade*. Available from: http://handle.dtic.mil/100.2/ADA053423 (reprinted with permission from Boeing Inc.)

## 5.7 Propeller efficiency ($\eta_p$)

Propeller efficiency $\eta_p$ is more of a dynamic expression, capturing performance changes in relation to the forward velocity of the aircraft. The expression for propeller efficiency uses the Advance Ratio, J (sometimes referred to as $\mu$ in helicopter analyses):

$$\text{Advance Ratio } (J) = \frac{\nu}{nD} \tag{5.16}$$

This is a ratio of the forward velocity, $v$ (m/s) to the propeller angular velocity, $n$ (rev/s) multiplied by the propeller diameter, $D$ (m). J is a dimensionless quantity which typically ranges between 0 and 2.

The Advance Ratio is then used to form the propeller efficiency equation:

$$\text{Propeller Efficiency, } \eta_p = J\frac{C_T}{C_P} \tag{5.17}$$

As stated previously, propeller efficiencies of 0.7–0.8 (70–80%) are considered to be excellent. Many scale props suffer from low efficiencies in the region of (0.3–0.5), this is mostly caused by operating in a low Reynolds number regime ($Re < 100{,}000$) and partly due to flapping caused by poor choice of low strength polymer materials. In larger aircraft, there is also the compressibility (Mach number) effect to be considered, whereby, the velocity at the tip approaches the speed of sound (340.29 m/s) causing shock wave effects. To compound this, there is also a reverse flow region in forward flight.

The figures below show typical wind tunnel test plots of $C_T$ & $C_P$ *vs J and* $\eta_p$ *vs J* for 10″ × 5″ and 10″ × 7″ APC propellers from the UUIC and OSU databases.[11]

As can be seen, the propeller efficiency changes with J and drops very quickly to zero once a maximum operating point is reached. Knowledge of the point of maximum efficiency can help to maximise the endurance of any type of aircraft which uses such scale propellers. As already mentioned, very few propeller manufacturers provide detailed real-world performance data[12] and therefore a whole industry of amateur enthusiasts who build their own test-rigs to perform this task has

11 Brandt, J.B., Deters, R.W., Ananda, G.K. & Selig, M.S. (1 January 2016) *UIUC Propeller Database*. University of Illinois at Urbana-Champaign. Available from: http://m-selig.ae.illinois.edu/props/propDB.html

12 APC (2016) *Computerised Data*. Available from: www.apcprop.com/v/PERFILES_WEB/listDatafiles.asp

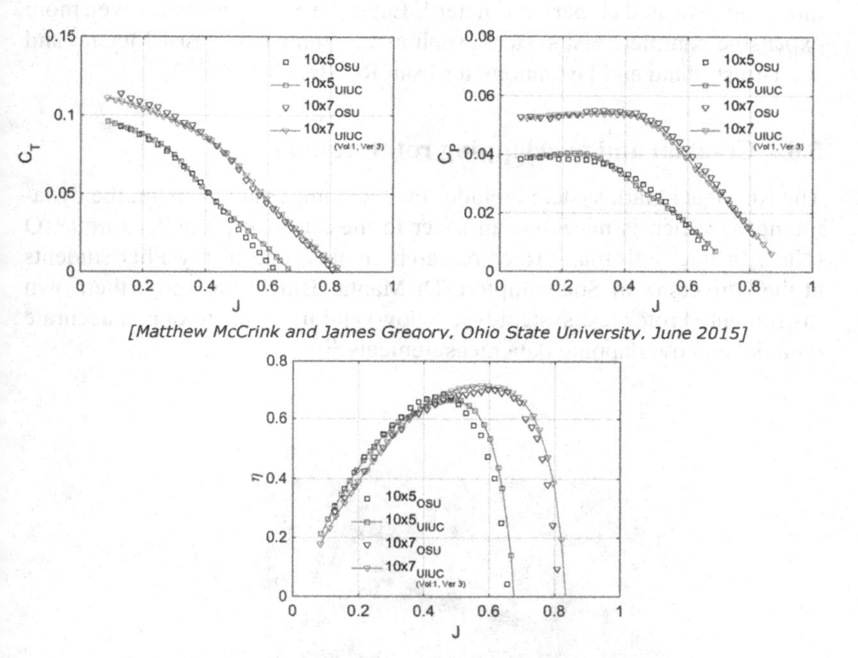


*Figure 5.15* Wind tunnel measurements of 10″ × 5″ and 10″ × 7″ APC propellers.

developed.[13,14,15] However, very few have access to instrumented wind tunnels to conduct accurate dynamic tests.

Other factors that can affect the efficiency, are the propeller diameter, pitch, angular velocity and airfoil profile. As already mentioned, the span-wise chord, taper and material choices also play their part.

Several tools are now available to help the researcher measure these quantities, and therefore predict the performance of their propellers, where data may not exist. These range from the relatively low-cost HobbyKing Wattmeter[16] that can measure the power of the propulsion system, to the

13 Fly Brushless (2018) *Propeller Search Tool*. Available from: http://flybrushless.com/prop/search

14 Empire RC (2018) *Prop Talk*. Available from: www.empirerc.com/hp-proptalk.htm

15 Empire RC (2018) *Prop Constants*. Available from: www.empirerc.com/hp-propconstants.htm

16 HobbyKing (2018) *Watt Meter*. Available from: https://hobbyking.com/en_us/turnigytm-7in1-mega-meter-battery-checker-watt-meter-servo-tester.html

more sophisticated Hyperion Emeter,[17] Eagle Tree eLogger[18] and even more expensive complete test systems such as the Tahmazo Thrust Meter[19] and the Thrust Stand and Dynamometer from RC Benchmark.[20]

## 5.8 Co-axial and overlapping rotor designs

The RC Benchmark system includes the most important element, the dynamometer, which is necessary in order to measure the propeller torque, Q (Nm). In line with many rotor research centres, one of my PhD students at the University of Southampton (Dr Mantas Brazinskas) built their own instrumented rotor test system (see below) and used this to obtain accurate co-axial and overlapping data measurements.[21]

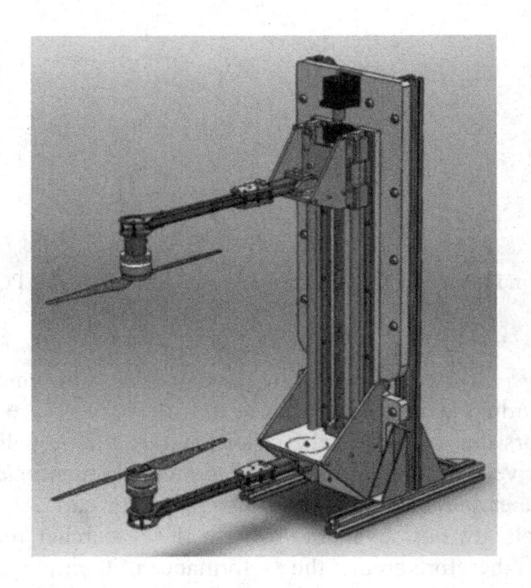

*Figure 5.16* Co-axial test-rig for overlapping rotor studies (Brazinskas *et al.*, 2016).

17 Empire RC (2018) *Emeter Data Logger*. Available from: www.empirerc.com/emeter.htm
18 Eagle Tree Systems (2018) *E-Logger*. Available from: www.eagletreesystems.com/index.php?route=product/product&product_id=54
19 Tahmazo (2018) *Thrust Stand*. Available from: www.tahmazo.com/products/catalogue/10/137
20 RCBenchmark (2018) *Dynamometer Series 1580*. Available from: www.rcbenchmark.com/dynamometer-series-1580/
21 National Instruments (2018) *Case study: Quantification of Overlapping Rotor Interference Effects of UAVs*. Available from: http://sine.ni.com/cs/app/doc/p/id/cs-17446

The purpose-built test-rig incorporated National Instruments sensors and linear actuators to automate the testing process using a range of both $z/D$ ratios, as well as $d/D$ ratios. This provided a complete understanding of the range of possibilities from completely separated rotors ($d/D > 1$) to fully co-axial ($d/D = 0$), as well as controlling the rotor-rotor vertical separation ($z/D$).

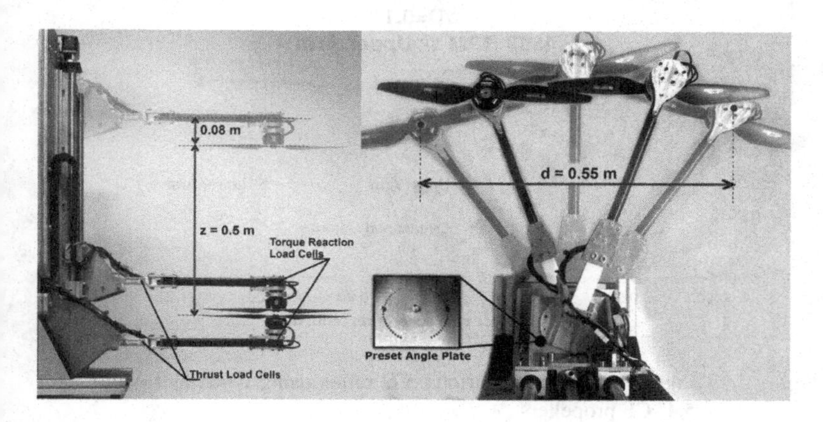


*Figure 5.17* Overlapping instrumented rotor test-rig using NI LABVIEW and CRIO.

This device has enabled the researcher to investigate and quantify the rotor-rotor interference effect of small rotors in the range of 15–18″.

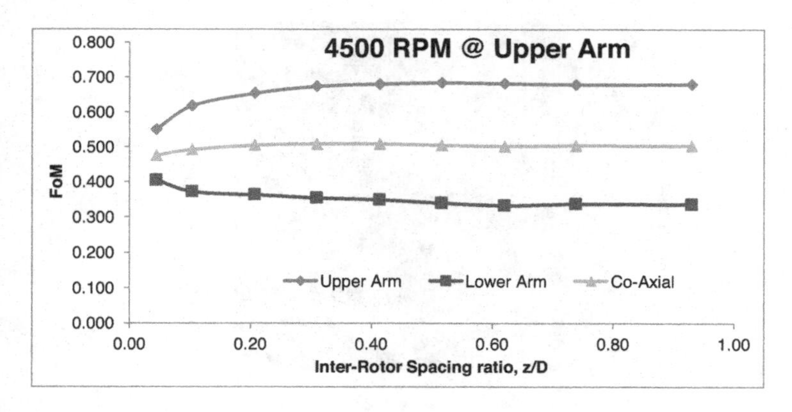


*Figure 5.18* Figure of Merit for various $z/D$ ratios using co-axial T-Motor 16″ × 5.4″ CF propellers.

Figure 5.18 shows that for this particular 16″ × 5.4″ CF propeller, as the $z/D$ ratio increases, the lower rotor FoM decreases, whereas the upper rotor

FoM increases. The overall effect is an increasing FoM with increasing *z*/*D* ratio up to about 0.2. There is little to be gained from increasing the *z*/*D* ratio beyond 0.2. Similar results have been found by Ramasamy (2015) at the UARC-NASA Ames Research Centre.[22]

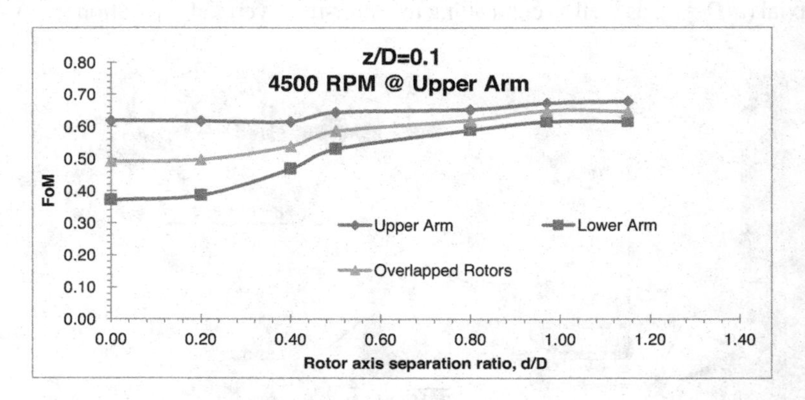


*Figure 5.19* Figure of Merit for various *d*/*D* ratios using co-axial T-Motor 16″ × 5.4″ CF propellers.

Figure 5.19 shows that there may be a small aerodynamic advantage to having rotors overlapping by 5–10%. This would also permit a smaller platform footprint.

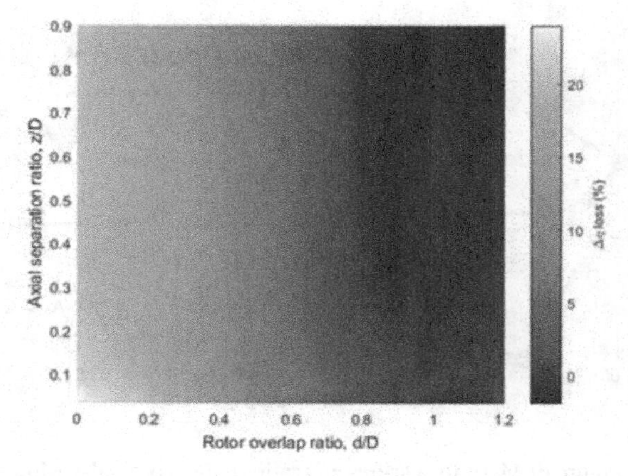


*Figure 5.20* Rotor-rotor interference losses *z*/*D* vs *d*/*D*.

22 Ramasamy, M. (2015) *Hover performance measurements toward understanding aerodynamic interference in coaxial, tandem, and tilt rotors*. Available from: www.ingentaconnect.com/contentone/ahs/jahs/2015/00000060/00000003/art00005?crawler=true

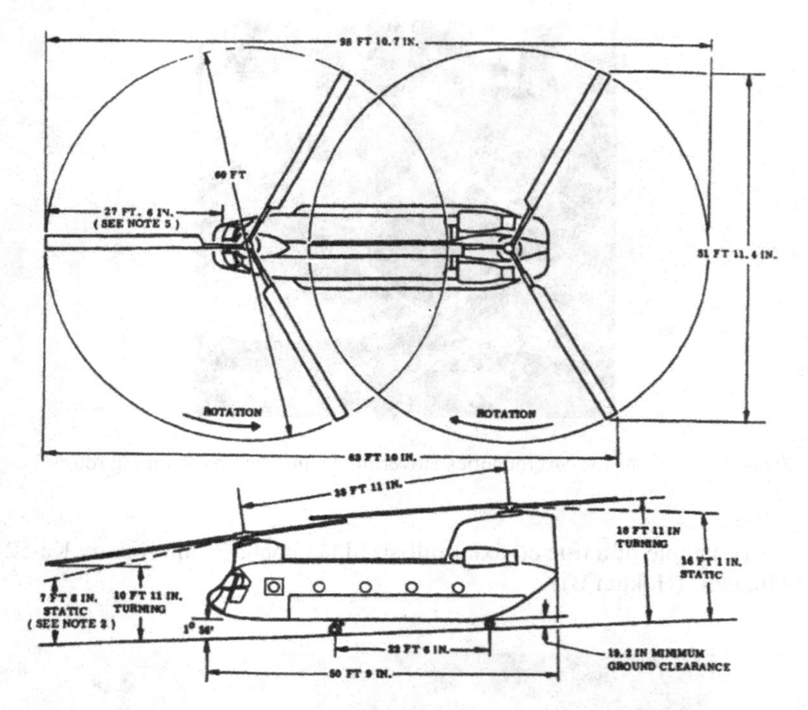


*Figure 5.21* The Boeing CH-47 Chinook Rotor dimensions showing the overlap (© Dover Publ. Inc. New York, USA).[23]

The Boeing CH-47 Chinook makes use of both of these features (overlapping and offset rotors) to good effect as can be seen. When calculated, the rotor-rotor overlap is 35% ($d/D = 0.65$) and the $z/D$ ratio is 0.15:

$$\text{Rotor Overlap} = 1 - (d/D) \tag{5.18}$$

The Malloy Aeronautics Hoverbike™ is another example of a small unmanned/manned aircraft making use of the overlapping rotor arrangement. It claims to carry a 130 kg payload at speeds of up to 60 mph.

23 Stepniewski, W. & Keys, C. (2003) *Rotary-Wing Aerodynamics*. Dover Publications Inc. New York, USA ISBN: 978-0486646473

*Figure 5.22* The Malloy Aeronautics Hoverbike™ utilising overlapping rotors.

An example of a true co-axial full-sized helicopter is the Kamov Ka-52 'Alligator' (Hokum B).

*Figure 5.23* The Kamov Ka-52 'Alligator' (Hokum B) co-axial military helicopter. Photograph by Fedor Leukhin.

The Kamov military helicopter range consists of all co-axial rotor designs, which operate at an average *z*/*D* ratio of 0.09 as can be seen below.[24]

24 Coleman, C.P. (1997) *NASA Technical Paper 3675, A Survey of Theoretical and Experimental Coaxial Rotor Aerodynamic Research*. Available from: https://ntrs.nasa.gov/archive/nasa/casi.ntrs.nasa.gov/19970015550.pdf

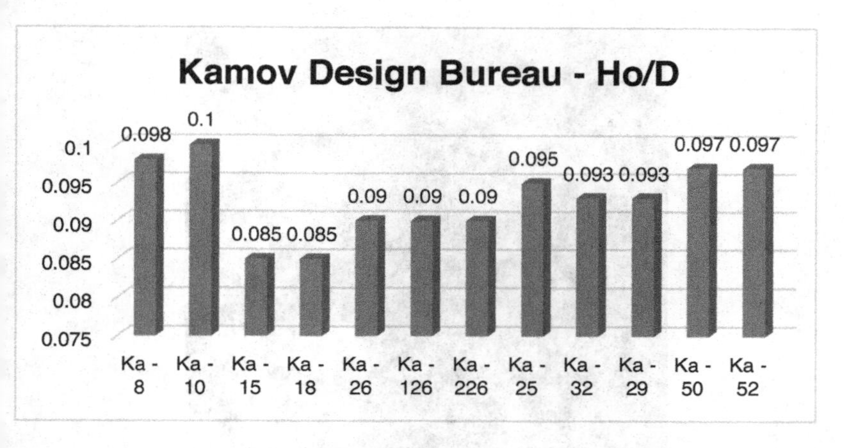


*Figure 5.24* $z/D$ ratio data from the Kamov range of military co-axial helicopters.

### *5.8.1 Different upper and lower rotors*

The standard co-axial configuration would consist of identical upper and lower rotors. However, several designs make use of a larger diameter lower rotor to increase the amount of 'clean' air not being interfered with by the prop wash of the upper rotor. Researchers have also argued for the increase in pitch of the lower rotor to enable it to cope with the increased inflow from the upper rotor. So, for example on the Cortex co-axial quad (X8) design from CarbonCore, the designer recommends 15″ × 5″ props for the upper rotor and 16″ × 5.4″ props for the lower rotor.[25]

## 5.9 Reynolds number, *Re*

One of the most fundamental problems of flying at scale is the low Reynolds number regime. The analogy of this is that the air becomes more viscous at this scale and therefore it is harder to fly efficiently.

The natural world of birds and insects has solved this problem by evolving flapping wing systems (Ornithopters) with incredibly low mass, in the milligram range. For example, a bee flaps its wings at 230 beats/s, can fly at 10 m/s and has a mass of only 100 mg.

25 CarbonCore (2018) *Cortex Multicopter*. Available from: www.carboncore.com/

*Figure 5.25* The Bee shows us the way to fly efficiently at scale. Photograph by Sffubs: https://commons.wikimedia.org/wiki/User:Sffubs).

The Reynolds number was pioneered by Stokes in the 1850s, but was named after the Irish engineer Osborne Reynolds in the 1880s for its use in pipe flow problems. The Reynolds number equation is the ratio of inertial forces to viscous forces:

$$Re = \frac{inertial\ forces}{viscous\ forces} = \frac{\rho VL}{\mu} = \frac{VL}{\nu} \tag{5.19}$$

Where:

$\rho$ *Air density (1.225 kg/m$^3$ at sea level, 15°C)*
$V$ *Velocity (m/s)*
$L$ *Characteristic length – for an airfoil this is the Chord length, c (m)*
$\mu$ *Dynamic viscosity of Air at 15°C (1.7965 x 10$^{-5}$ Pa s)*
$\nu$ *Kinematic viscosity of Air at 15°C (1.4657 x 10$^{-5}$ m$^2$/s)*

The figure below puts the importance of Reynolds number into the context of things that fly, from commercial airliners to birds and insects. The blue shaded area on the graph contains the operational range of micro air vehicles (MAVs). Note that these all typically fly below 100,000 *Re*.

The importance of 100,000 *Re* cannot be overstated, since it is at this level that an important transition takes place between the Lift to Drag ratio, as can be seen in the figure below.

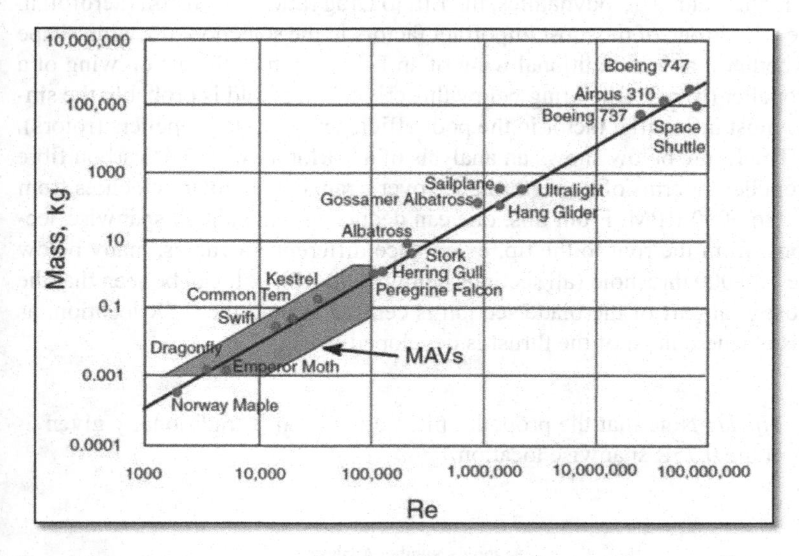


*Figure 5.26* Relationship between flyer body mass and Reynolds number.[26]

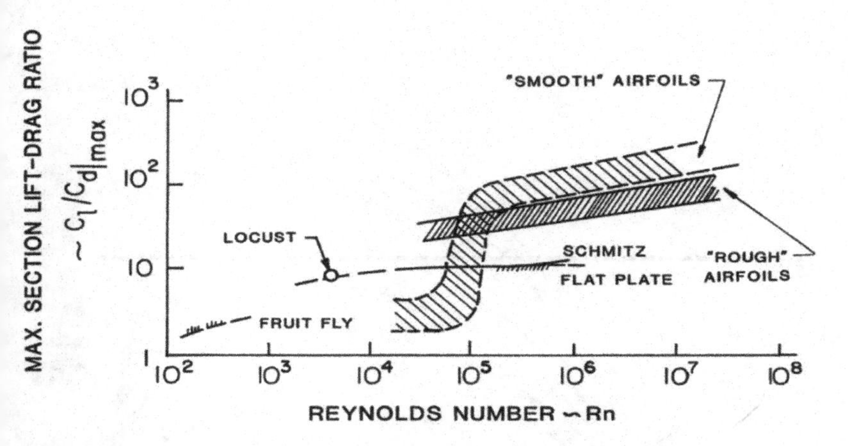


*Figure 5.27* Transition region for lift/drag ratio at 100,000 *Re*.[27]

26 Bohorquez, F. (2007) *Rotor Hover Performance and System Design of an Efficient Coaxial Rotary Wing Micro Air Vehicle*. PhD thesis, University of Maryland, College Park, Department of Aerospace Engineering.

27 McMasters, J. H. & Henderson, M. L. (1980) Low-Speed Single-Element Airfoil Synthesis. Available from: http://journals.sfu.ca/ts/index.php/ts/article/view/989/943

In the field of aerodynamics, the Lift to Drag ratio of an airfoil (aerofoil in the UK) is one of the most important factors in the selection of a wing shape (whether that is a traditional wing of an FW aircraft or the airfoil wing of a propeller (rotor). Operating below this critical threshold is probably the single most influential factor in the poor efficiency of scale propellers (rotors).

The figure below shows an analysis of a T-Motor 16″ × 5.4″ carbon fibre propeller in terms of its performance over a range of angular velocities from 500 to 6000 RPM. From this, one can deduce that the airfoil spanwise sections, from the root to the tip, experience different *Re* values, many below the 100,000 threshold (all sections below 3000 RPM). It can be seen that the most vital part of the blade section is centred around the 0.7R location, as this is where most of the thrust is developed.

***Tip 11:*** Note that the propeller pitch dimension is traditionally given at the 0.75R spanwise location.

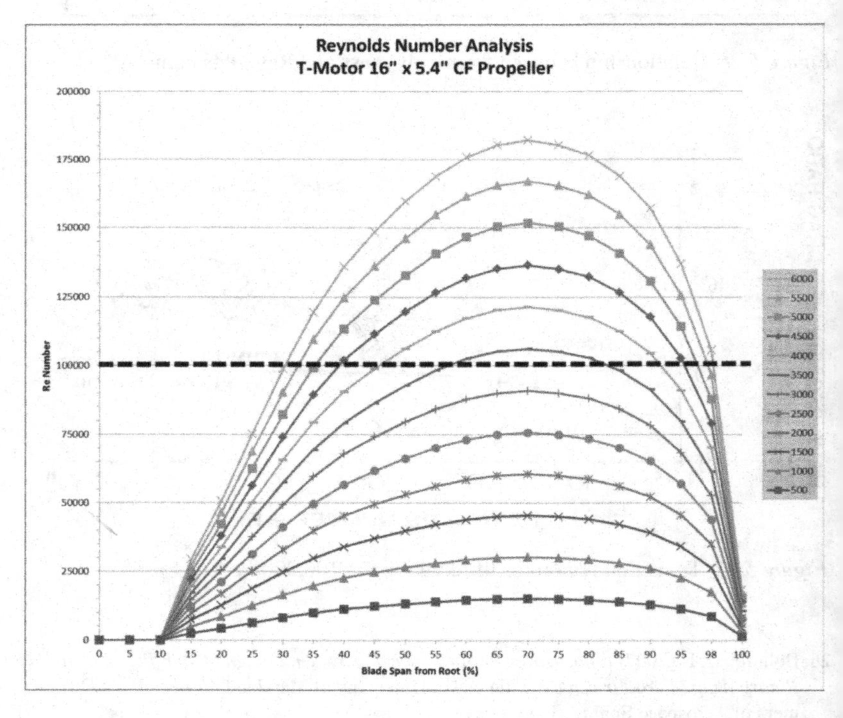


*Figure 5.28* Analysis of spanwise Reynolds number variation for a T-Motor 16″ × 5.4″ CF propeller.

## 5.10 Airfoil geometry

Whereas most people will not know what airfoil is used in their aircraft wing or propeller (rotor), or even be in a position to choose one given the option, it is still important to understand the fundamentals. For an easy to understand introduction to airfoil design and use, the reader is directed to the excellent NASA website where you can download their FoilSim III applet.[28]

There are several thousand airfoil geometries available to the aircraft designer, some of these specifically designed for low-Reynolds number environments.[29]

A generic airfoil geometry is presented in the figure below.

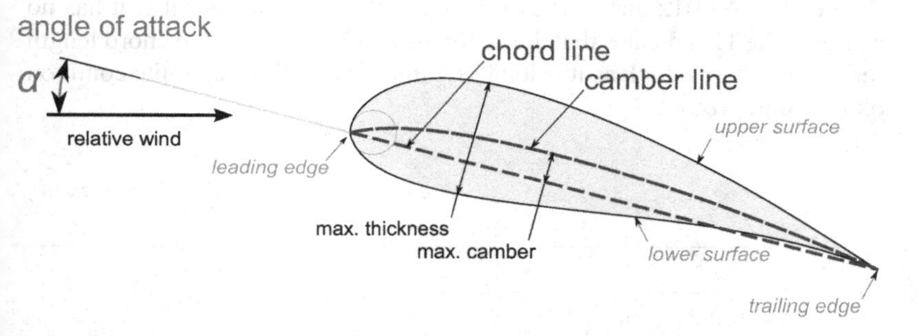


*Figure 5.29* Generic airfoil geometry.

There is often a misconception of how an airfoil actually works. XFOIL is an interactive program for the design and analysis of subsonic isolated airfoils. It consists of a collection of menu-driven routines which perform various useful functions.[30]

The NACA airfoil series are airfoil shapes for aircraft wings/props developed by the National Advisory Committee for Aeronautics (NACA) in the first half of the 20th century. The profile of the NACA airfoils is described using a series of digits following the word 'NACA'.

28 NASA (2018) *FoilSim III – Airfoil Simulator*. Available from: www.grc.nasa.gov/WWW/K-12/airplane/foil3.html

29 Selig (2018) *UUIC Low-speed Airfoil Tests*. Available from: http://m-selig.ae.illinois.edu/uiuc_lsat.html

30 Drela (2018) *XFoil Subsonic Airfoil Development System*. Available from: http://web.mit.edu/drela/Public/web/xfoil/

The parameters in the numerical code can be entered into equations to precisely generate the cross-section of the airfoil and calculate its properties. For example, the NACA 2412 airfoil has a maximum camber of 2% located 40% (0.4 chord) from the leading edge with a maximum thickness of 12% of the chord.

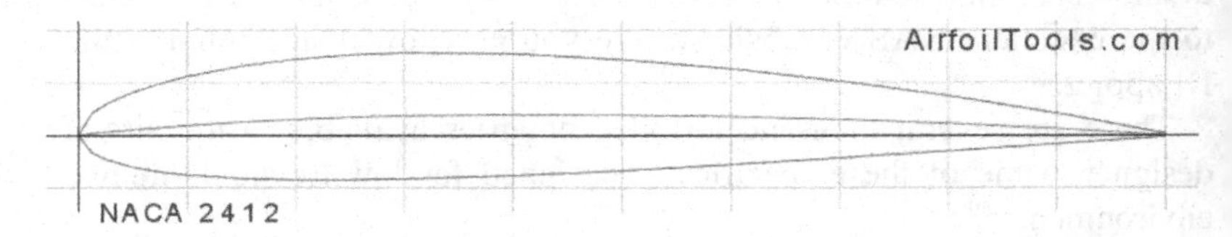


*Figure 5.30* NACA 2412 airfoil.

The NACA 0012 airfoil is symmetrical, the 00 indicating that it has no camber. The 12 indicates that the airfoil has a 12% thickness to chord length ratio: it is 12% as thick as it is long. Symmetrical airfoils are quite common on helicopter rotors.

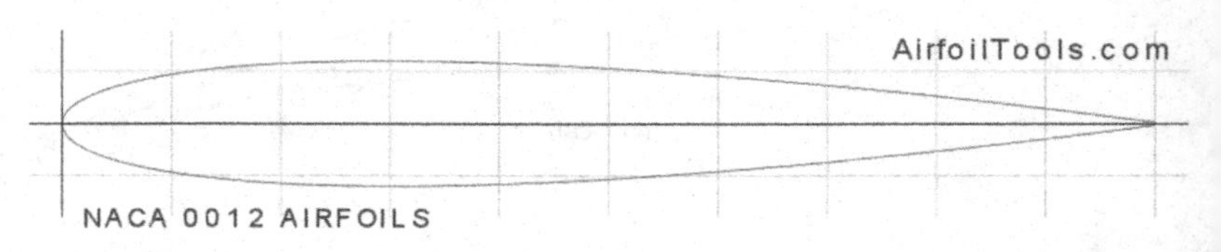


*Figure 5.31* NACA 0012 symmetrical airfoil.

*Figure 5.32* Westland Scout helicopter showing rotor airfoil geometry.

## 5.11 Airfoil theory – Blade Element Theory (BET)

To understand the basic theory of how an airfoil produces lift (and drag) it is important to investigate the underlying forces working at the aerodynamic interface. To do this, we will utilise a method called the Blade Element Theory (BET). In Section 5.4, the Simple Momentum Theory (SMT) introduced a crude overview of the system from a streamtube perspective, without considering the propeller (rotor) element. It is now time to look at this important component in greater detail.

The first thing to consider when looking at a propeller (rotor) system is the fact that as well as the rotating airfoil (tangential) velocity, $W_t$, we also have a forward (axial) velocity, $W_a$. The resultant is the velocity vector, $W$.

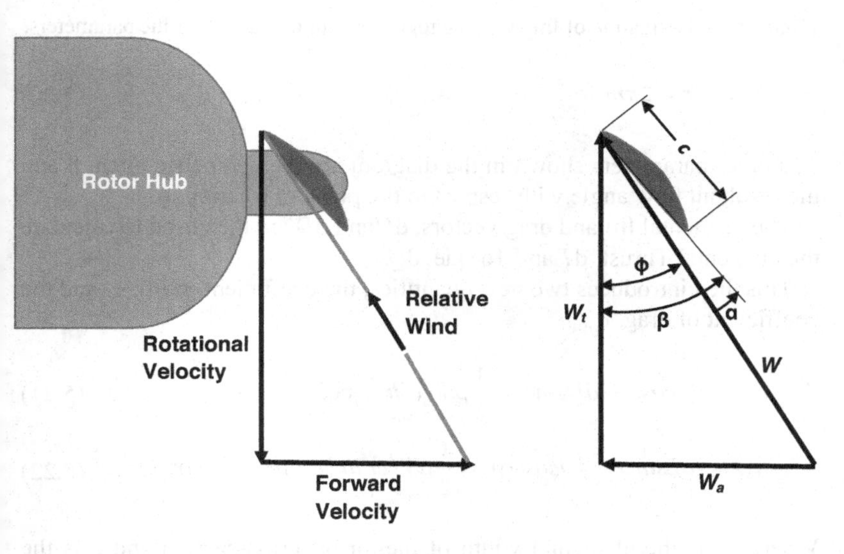


*Figure 5.33* Analysis of propeller relative velocity term.[31]

Here we will define the angle of attack, $\alpha$ as the angle from the resultant vector, $W$ and the chord line of the airfoil. Note that the drag, $dD$ is parallel to this resultant vector, $W$ and that the lift, $dL$ is perpendicular to this. Because this airfoil cross-section represents a small slice of the whole

31 Curtiss-Wright Corporation (1944) *Propeller Theory*. Propeller Division, Caldwell, NJ, USA, pp. 37.

propeller (at some arbitrary radius, *r*), the axial velocity, $W_t$ increases from the root to the tip:

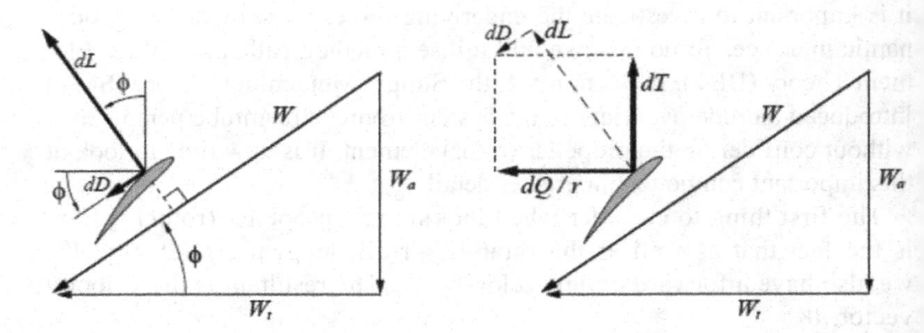


*Figure 5.34* Derivation of lift and drag together with thrust and torque parameters.

$$W_t = \Omega r = 2\pi nr \tag{5.20}$$

The other parameters shown in the diagram are the geometric pitch, β and the resultant flow angle with respect to the plane of rotation, ϕ.

The elemental lift and drag vectors, d*L* and d*D* are then used to calculate the elemental Thrust, d*T* and Torque, d*Q*/*r*.

This then introduces two new quantities, the coefficient of lift, $c_l$, and the coefficient of drag, $c_d$:

$$dT = dLcos\phi - dDsin\phi = \frac{1}{2}\rho W^2 c\, dr\left(c_l\cos\phi - c_d sin\phi\right) \tag{5.21}$$

$$dQ = \left(dLsin\phi - dDcos\phi\right)r = \frac{1}{2}\rho W^2 cr\, dr\left(c_l\sin\phi - c_d cos\phi\right) \tag{5.22}$$

Where d*r* is the elemental width of the airfoil cross-section and *c* is the airfoil chord.

By integrating over the entire radius, *R* of the propeller (rotor), the total thrust, *T* (N) and the total torque *Q* (Nm) can be calcuated.

## 5.12 Online airfoil resources

An excellent airfoil resource (containing 1636 airfoils) of common profiles is the Airfoil Tools website.[32] From this dataset, a designer can select,

32 Airfoil Tools (2018) *Airfoil Geometry Visualiser*. Available from: http://airfoiltools.com/airfoil/details?airfoil=naca2412-il

compare and contrast airfoil designs in terms of their performance against a range of Reynolds numbers.

As a design exercise, the database was used to search for the highest lift/drag ratio at a range of Reynolds numbers from 50,000 to 500,000 (see Table 5.1).

*Table 5.1* Variation of the airfoil maximum lift/drag ratio with Reynolds number.

| *Reynolds Number, Re* | *Max CL/Cd Ratio* | *Airfoil Name* |
|---|---|---|
| 50,000 | 59.2 | (Gottingen 79) |
| 100,000 | 105.6 | (Eppler 376) |
| 200,000 | 162.0 | (Gottingen 448) |
| 500,000 | 176.8 | (Eppler 63) |

Given that the Reynolds number changes with the angular velocity of the airfoil; in designing a propeller (rotor) the aerodynamicist is always searching for a tolerant airfoil which maintains a high lift/drag ratio over a wide range of Reynolds numbers.

The Eppler 63 airfoil (as used in the APC multi-rotor range) is such a design, as can be seen from the following polar diagrams (each colour represents a different Reynolds number (50,000–500,000)). These polar plots give the designer a clear indication of the likely performance of the airfoil against various conditions and angles of attack (Alpha).

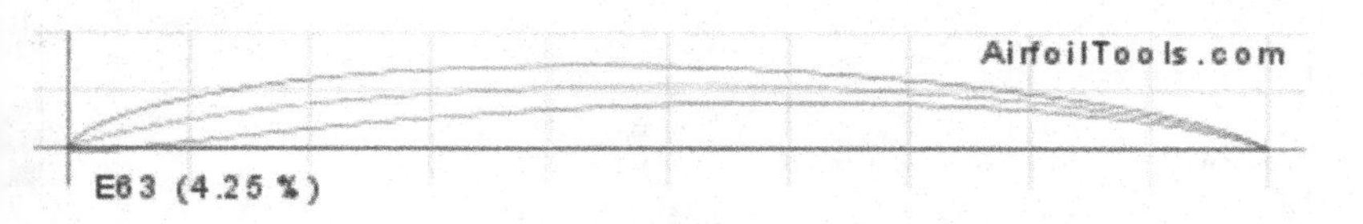


*Figure 5.35* Eppler E63 low Reynolds number airfoil.

As can be seen, the angle of attack (Alpha) can have a dramatic effect on the performance of the airfoil; too much Alpha and the airfoil will begin to stall. The optimum angle of attack (Alpha) for this airfoil would be somewhere between 1° and 5°, depending on the local Reynolds number. If this airfoil proved too difficult to manufacture due to its thin section, the designer might look at other slightly thicker airfoil sections such as the Archer A18 (smoothed) as shown below.

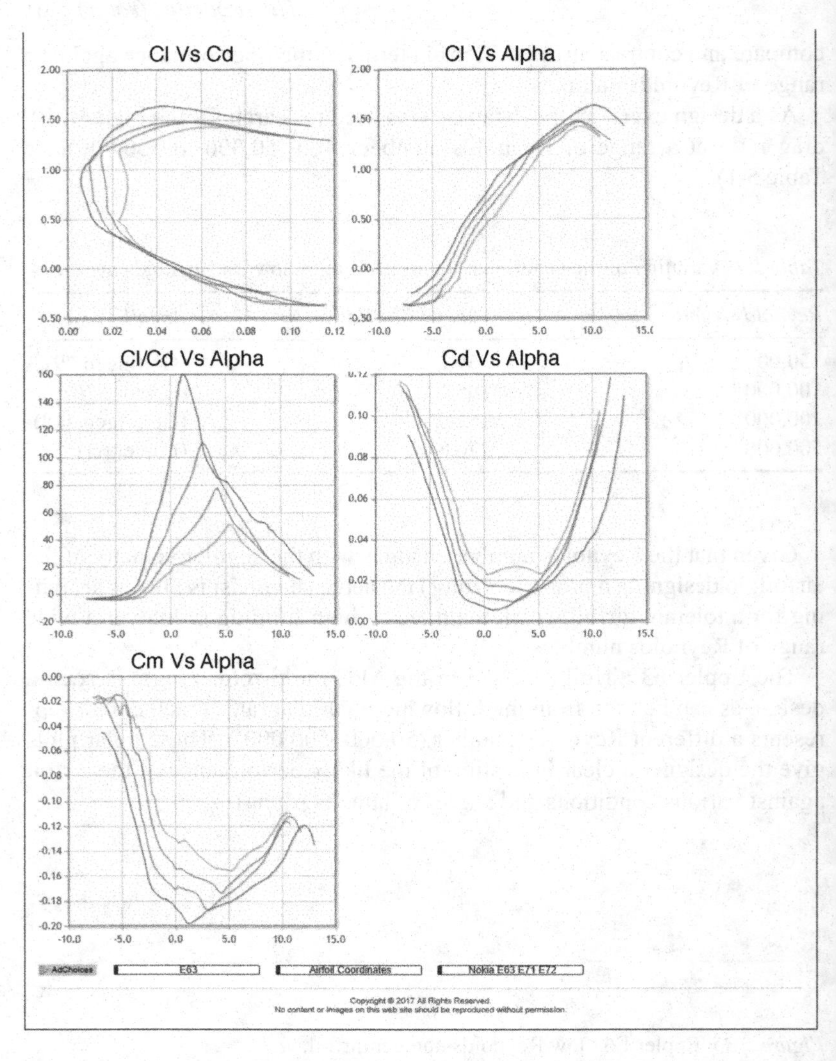


*Figure 5.36* Polar plots of an E63 low Reynolds number airfoil (*Re* = 50,000–500,000).

*Figure 5.37* Archer A18 low Reynolds number airfoil.

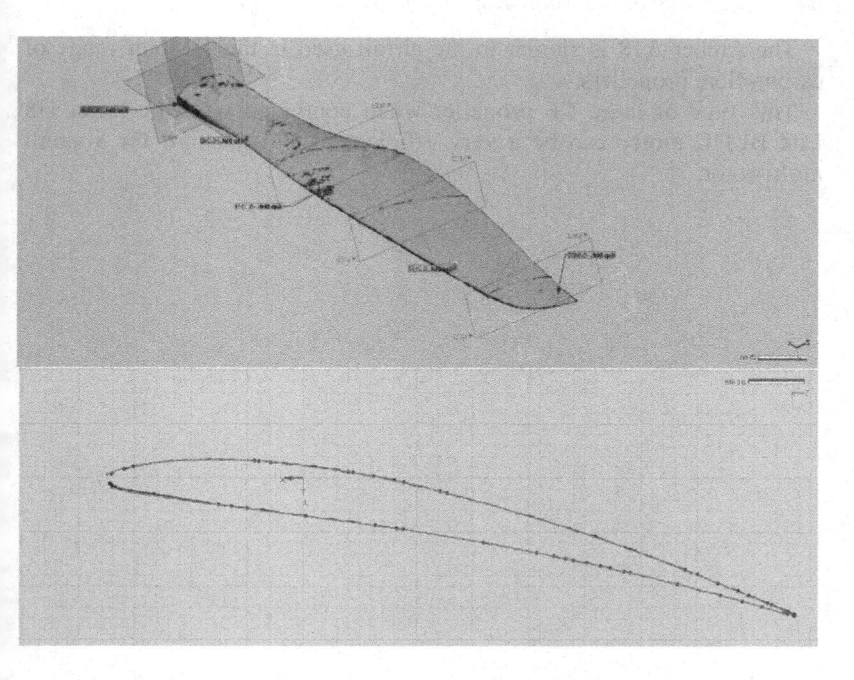

*Figure 5.38* Reverse engineered 3D scan of a 29.2″ × 9.5″ T-Motor CF propeller airfoil section.

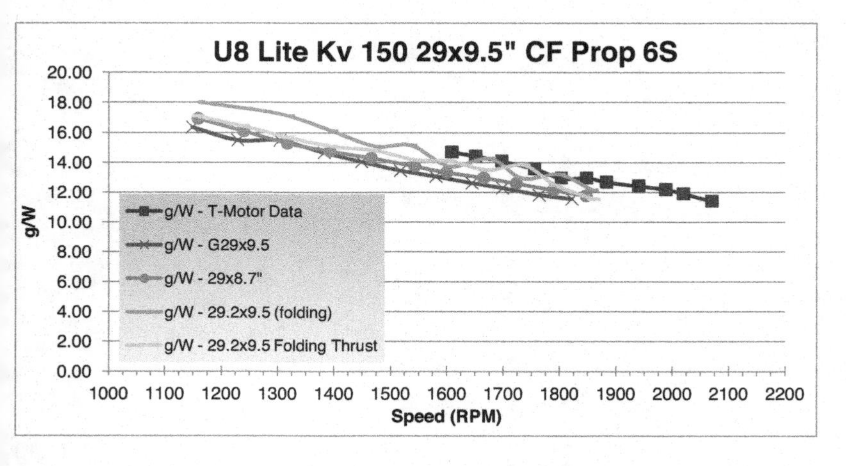


*Figure 5.39* Efficiency metric (g/W) curves for various 29″ T-Motor CF propellers.

The Archer A18 is similar to the airfoil used in the T-Motor range of carbon fibre propellers.

This type of large CF propeller when combined with the latest U8 Lite BLDC motor can be a very efficient propulsion unit for a small multi-rotor.

# 6 The system solution

Having discussed all the individual components in some detail, it is now appropriate to discuss the overall system solution incorporating all the knowledge acquired so far.

As stated in Chapter 1, the starting point of any design is usually either the customer user requirements or the competition rules. For the IMechE UAS Challenge, the starting point is always the rules, which are released in October in the year before the competition fly-off in June.

This design process is somewhat iterative in nature, with the designer/engineer cycling backwards and forwards around the design circle, even completing multiple circuits before honing in on an optimum design selection.

Given the initial premise that we are working on the design of a fully electric multi-rotor, the main user requirements will consist of the big five: Endurance, Speed, Payload Capability, MTOM and Cost. Of these five, the IMechE UAS Rules stipulated that the MTOM must be less than 7 kg and that the total cost of COTS must be no more than £ 1000. Teams are encouraged to develop solutions with high payload capability of up to 3 kg (a mix of either 1 kg or 0.5 kg flour bags). Although this was optional, more points could be gained by carrying more mass. Likewise, the endurance challenge (number of laps) was actually more of a speed challenge, since the total endurance was limited to a maximum of 10 min for each flight operation. At the furthest waypoint, communications can become an issue (see Appendix F).

The 2017 challenge consisted of three main elements: [1] A payload delivery, [2] An endurance challenge and [3] A reconnaissance mission. There was a total of 500 points available (250 of which could be gained during the fly-off demonstration).

An analysis of the rules provided teams with a clear path to success; highlighting the importance of maximising payload and having a fully autonomous system.

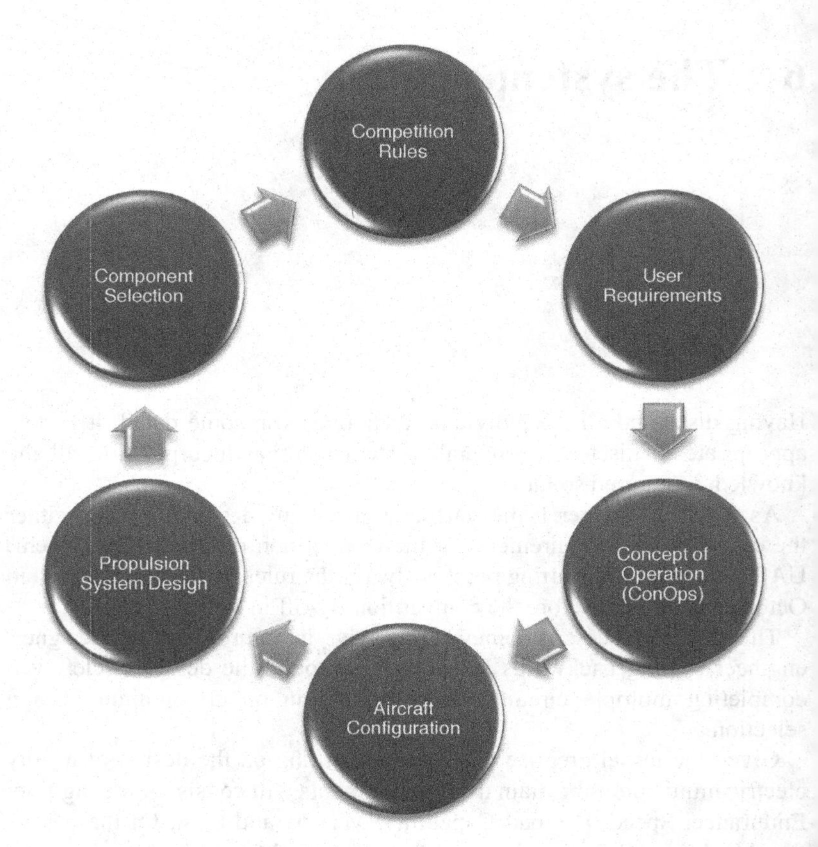


*Figure 6.1* Multi-rotor system design wheel.

*Table 6.1* Submission breakdown (IMechE UAS Challenge Rules 2017).

| *Element* | *Sub-element* | *Points* |
|---|---|---|
| Design | PDR Submission | 25 |
| Design | CDR Submission | 85 |
| Design | Dragon's Den Business Case Presentation | 25 |
| Design | Environmental Impact Poster | 15 |
| Flight Readiness | Design Presentation and FRR Submission | 50 |
| Flight Readiness | Scrutineering and Manufacturing Poster | 50 |
| Flight Demonstration | Mission 1 – Payload Delivery | 100 |
| Flight Demonstration | Mission 2 – Reconnaissance | 75 |
| Flight Demonstration | Mission 3 – Endurance | 75 |
| **TOTAL** | | **500** |

*Table 6.2* Analysis of Scoring Priorities (IMechE UAS Challenge Rules 2017).

| *Success category* | *Total points* |
|---|---|
| Maximise payload | 66 |
| Fully autonomous flight | 60 |
| Accurate target location | 32 |
| Autonomous alphanumeric recognition | 24 |
| Accurate navigation | 20 |
| Accurate payload delivery | 18 |
| Maximise battery capacity | 18 |
| Maximise speed | 12 |
| **TOTAL** | **250** |

Several elements listed earlier are beyond the scope of the discussion of an optimum multi-rotor propulsion system design. Some of these relate to the choice of autonomous flight controller (Pixhawk[1] in this case – see Appendix G) and the design and development of a target recognition system (Odriod[2] single board computer in this case).

## 6.1 IMechE UAS challenge – a worked multi-rotor example

This brings us back to the fundamental design and selection of the optimum propulsion system components for a small multi-rotor system of sub-7 kg, which has to carry a 3 kg payload. So this sets the complete system mass as no more than 4 kg. This is achievable, but is not an easy task.

By setting the MTOM at 7 kg, this dictates the total thrust requirement at the take-off point, which is 68.67/n (N) @ a throttle setting of 50% (please refer back to the design constraints on p. iii), where n equals the number of BLDC motors/rotors chosen for the aircraft configuration. We are therefore interested in a BLDC motor that can deliver this specification at a reasonable mass, cost and efficiency.

Clearly, there are an almost infinite number of possible selections available to the designer/engineer. One could look to the industry leaders (AXI, Hacker, Aveox and Plettenberg) to provide a solution or rely on a previously used motor manufacturer.

1 Pixhawk (2017) *Pixhawk Autopilot*. Available from: www.unmannedtechshop.co.uk/unmanned-pixhawk-autopilot-kit/

2 Odroid (2017) *Odroid Single-Board Computer*. Available from: www.odroid.co.uk/hardkernel-odroid-c2-board

A manufacturer of such systems, which has been mentioned many times in this book, is T-Motors, who are based in Southern China. The quality of their (German designed) BLDC motors, Props and ESCs is first rate, and these have been used successfully for many such platforms.

To help in the selection of the optimum T-Motor BLDC motor for the previous specification, a database of all the T-Motor BLDC motors was created in Excel. This enables the user to search through the entire catalogue within milliseconds to find the motor with the highest (g/W) efficiency metric at a particular design thrust (N), voltage (6–12S) and % throttle (50–100%).

An example of the search screen for this application can be seen in Appendix C. For this analysis, I have assumed that the aircraft configuration will be a hexrotor (n = 6) operating at 6S. As you can see from the search results, a number of BLDC motors (in order of highest to lowest g/W) are listed. This enables the designer/engineer to investigate other properties, such as motor mass, cost and likely prop combinations. The variance button allows the user to input a thrust variance above and below the chosen thrust condition to capture motors that otherwise might be lost due to not quite meeting the selection metric.

As can be seen from the results, the (g/W) efficiency metric ranges from 21 to 10.4 g/W, over the first 21 motors. Remembering that we are looking for an efficiency of about 10 g/W at take-off and given the need to fit into the 4 kg mass target for the platform, a good choice might be the U5 (kV 400) BLDC motor in combination with 16″ × 5.4″ CF props.

This choice would yield a total motor mass of 6 × 156 g = 936 g, a total prop mass of 6 × 28.2 = 169.2 g and an overall propulsion mass of 1.1 kg. If we consider that the chassis will come in at about 1 kg and the Li-Po battery will be about 1.3 kg (see Appendix A2), then we have just under 600 g for everything else (ESCs, FC, cables, camera, RX, etc.), which is feasible.

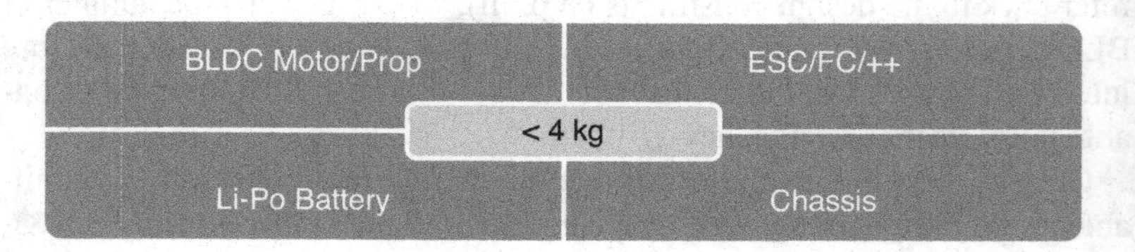


*Figure 6.2* System design breakdown and allocation of mass.

The process is then one of iteration, involving examining the performance of each individual component to ensure that they are all optimised. For example, in selecting a Li-Po battery, we would have to consider the likely mass, capacity, 'C' rating, voltage, etc. and its implication to the endurance.

To maximise endurance, we are searching for a Li-Po battery with a Specific Energy of around 200 Wh/kg at 6S. Let us therefore consider the MaxAmps 11,000 mAh 6S Li-Po due to its mass match (1270 g) and SE of 192 Wh/kg. Since this has a 'C' rating of 40 C, 440 A could be drawn continuously, which is more than enough for this application.

At hover, the U5 (kV 400) BLDC motor draws about 5.2 A, meaning that the total current draw would be approximately 6 × 5.2 A = 31.2 A. Observation of the T-Motor websites informs the user that this motor when combined with the 16″ × 5.4″ CF propeller will give a maximum thrust of 2850 g/motor (@ 100% throttle) for a current of 20 A (Total current draw 120 A). This provides a Thrust to Weight ratio of 2.44:1, which is within our design criteria set on p. iii.

This would also provide an endurance at hover (3 kg payload) of ((11 × 0.85 Ah)/31.2 A) × 60 = 18 min. (Note: the previous endurance estimate includes the 85% battery DoD condition).

Given that the total mission time set by the IMechE Rules is 10 min this would provide a safety margin of 1.8 for unforeseen circumstances, such as adverse wind, temperature or increased speed. A spreadsheet used to calculate these system permutations is shown in Appendix D.

Having conducted the previous optimisation exercise, the next logical step would be to purchase a single BLDC motor, CF Prop and ESC to run in-house static/dynamic thrust tests using either a purpose-built or commercial test-rig such as the RC Benchmark equipment mentioned in Chapter 5.

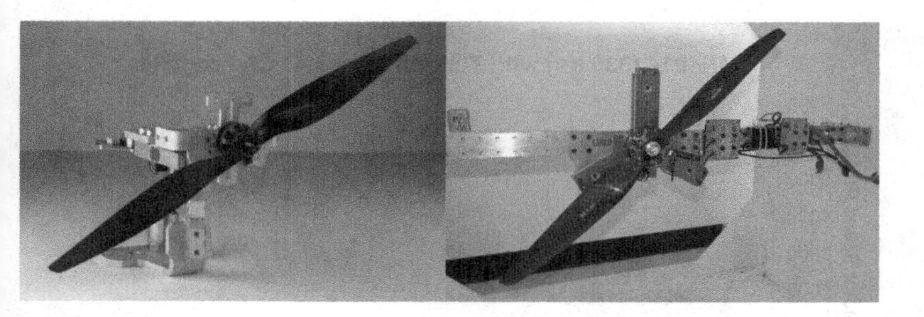

*Figure 6.3* RC Benchmark propeller test-rigs (Series 1580 (L) & 1780 (R)).

As stated previously, the benefit of these test-rigs are that they can measure all of the physical quantities of interest (throttle PWM, thrust, torque, speed, current, voltage) in real-time. Thus, it is a simple matter to configure a script to calculate the input power, output power, coefficient of thrust and power and from this the motor efficiency and Figure of Merit (FM).

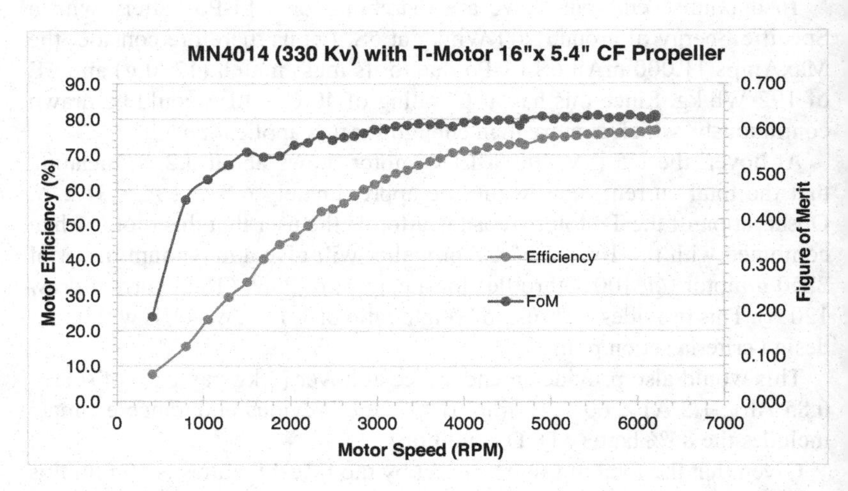


*Figure 6.4* Data from an RC Benchmark test showing the motor efficiency and FoM vs motor speed.

As shown in Figure 6.4, this gives the user all the information about the propulsion system and the optimum operating point. We can see that for this particular setup the motor will reach an efficiency of approximately 76% at 6000 RPM. Also at this point, the Figure of Merit (FM) has reached 0.62, which is quite good for such a scale propeller.

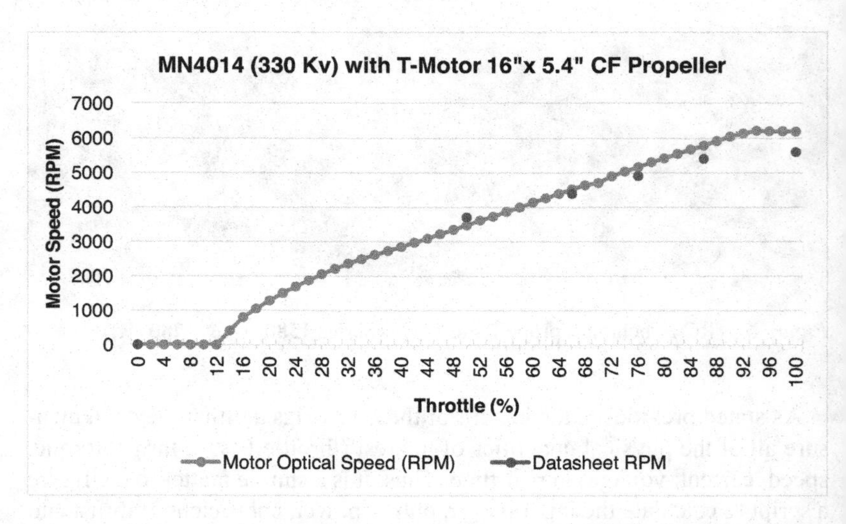


*Figure 6.5* The % throttle linearity with motor speed test using RC Benchmark test-rig.

## 6.2 Measuring the propulsive efficiency

The propeller efficiency equations from Chapter 5.4 can be used to calculate the FoM (Hover) and/or the propulsive efficiency. The RC Benchmark test-rig can even be mounted into a wind tunnel so that the propeller efficiency can be measured.

*Figure 6.6* The RJ Mitchell wind tunnel complex used to test our propellers.

As shown in the Figure 6.3, the RC Benchmark dynamometer has been used to measure the dynamic performance of a range of scale CF propellers (18–29″). This has also been utilised to measure the propeller (rotor) performance when the test-rig has been rotated through a range of pitch angles to simulate the oblique flight envelope (from traditional fixed-wing to rotary-wing at hover (0–90°)). This information is essential when designing cross-over (hybrid) aircraft configurations, such as the tilt-rotor V-22 Osprey.

*Figure 6.7* V22 Osprey tilt-rotor aircraft.

When viewed from the front of the aircraft, the propeller (rotor) planform changes from a circle through various conical sections (ellipses) ending at a singularity, known as a degenerate ellipse (line).

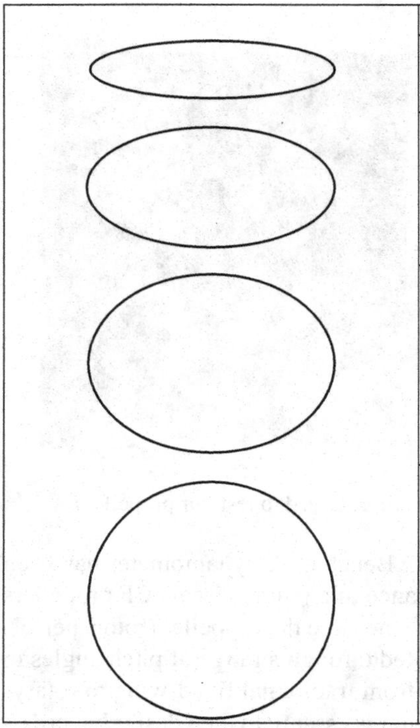

| | Tilt Angle = 90 deg | Projected Area = 0% |
|---|---|---|
| | 60 deg | 50% |
| | 45 deg | 71% |
| | 30 deg | 87% |
| | 0 deg | 100% |

*Figure 6.8* Analysis of a tilt-rotor geometry aircraft when viewed from the front.

When calculating the propeller efficiency using the Thrust coefficient, Power coefficient and Advance Ratio equations, care should be taken to resolve the forward velocity term to coincide with the tilt angle of the rotor disk.

Since most multi-rotor aircraft fly in the (45–90°) tilt angle range, this is the area of most interest to engineers and designers. As can be seen from Figure 6.8, at a tilt angle of 45°, the projected area is 71% of that of the full circle at 0°.

The relationship between the tilt angle and the projected area is non-linear (Cosine function) and can be seen in the figure below.

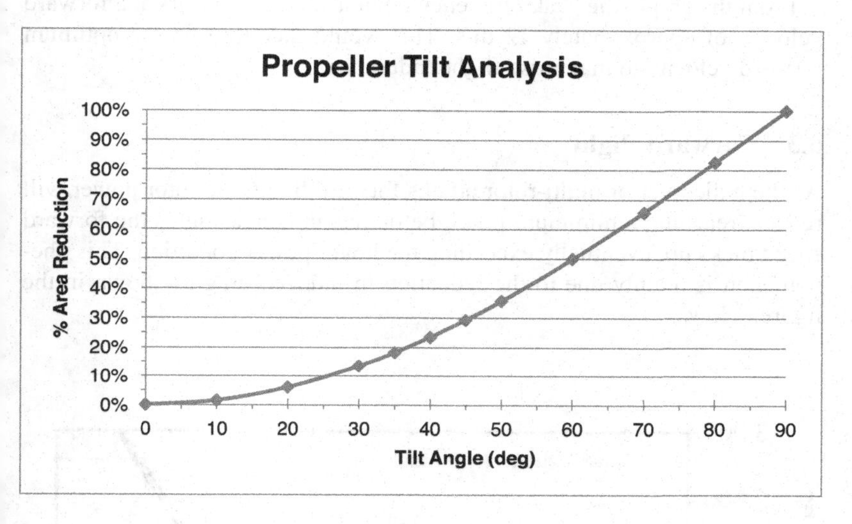


*Figure 6.9* Graph of rotor % area reduction against tilt angle.

Wind tunnel tests conducted in the large RJ Mitchell test facility at the University of Southampton's Highfield Campus by MSc students (Sampath & Whiteley), shows that there could be a small aerodynamic advantage (5%) to operating a propeller at a tilt angle in common with all multi-rotor configurations (see below).

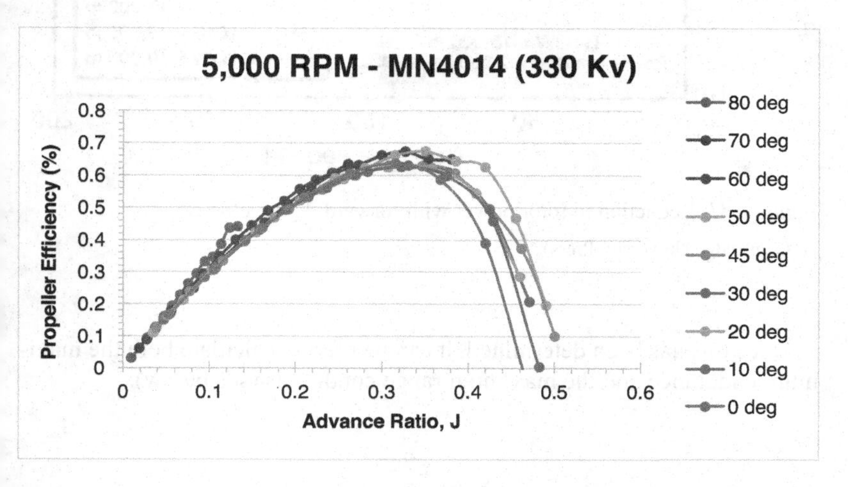


*Figure 6.10* Wind tunnel data for a T-Motor 18″ × 6.1″ CF propeller operated at various tilt angles.

From the graph, the peak efficiency point at $J = 0.32$ occurs at a forward velocity of approximately 12 m/s. This would therefore be the optimum forward velocity to maximise flight endurance.

## 6.3 Forward flight

As the helicopter or multi-rotor attains forward flight, the rotor power will reduce, reaching a minimum point, before again increasing as the forward speed picks up, eventually exceeding the hover power condition. This phenomenon is mainly due to the reduction in induced drag as shown in the figure below.

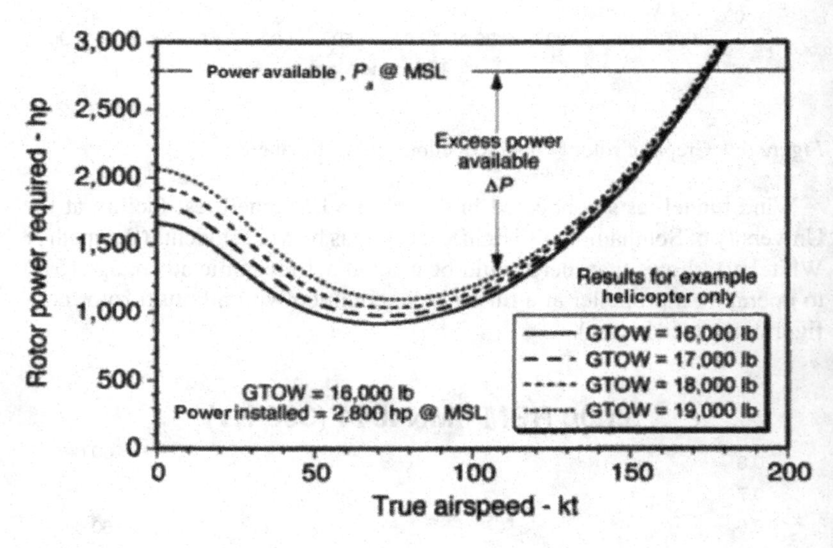


*Figure 6.11* Reduction in rotor power with forward flight velocity (© Cambridge University Press).[3]

Once this has been determined, it can be used to calculate both the maximum endurance and the maximum range conditions (see below).

3 Leishman, J.G. (2016) *Principles of Helicopter Aerodynamics*, 2nd edition. Cambridge University Press, Cambridge.

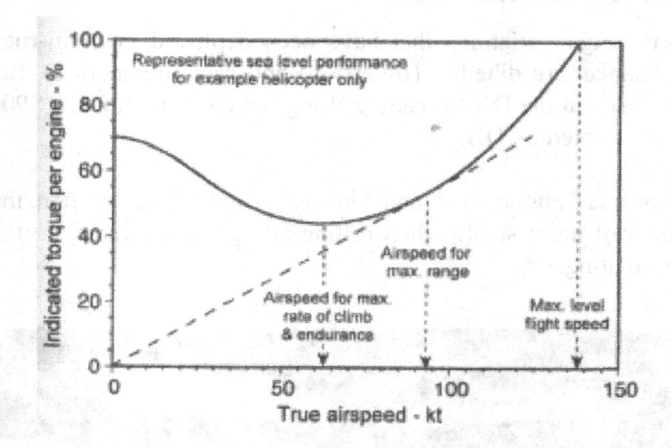


*Figure 6.12* Maximum endurance and range calculations
(© Cambridge University Press).[3]

In a small multi-rotor, this effect is much less pronounced, due to the reduced drag and smaller scale. A test conducted by Jake Ware, a PhD student, based at MIT in the US shows this effect (see below).

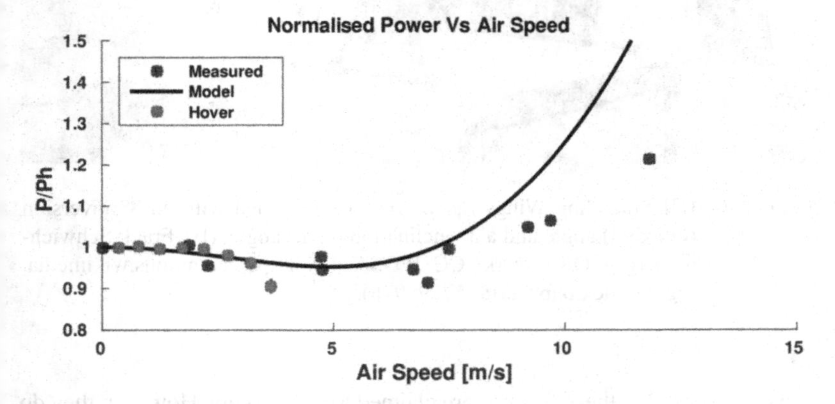


*Figure 6.13* Power curve for a small multi-rotor.[4,5]

4 Ware, J. & Roy, N. (2016) An analysis of wind field estimation and exploitation for quadrotor flight in the urban canopy layer. *ICAR'16 – International Conference in Advanced Robotics*, 16–21 May 2016, Stockholm, Sweden.

5 Wilkerson, S., et al. (2016). Aerial swarms as asymmetric threats. *2016 International Conference on Unmanned Aircraft Systems (ICUAS)*, 7–10 June 2016, Arlington, VA, USA.

## 6.4 Dihedral and cant angles

Two other design variations that have been deployed in multi-rotors to aid performance, are dihedral (Inversion) and cant (inclination). Both of these were used in the DJI Spreading Wings series of platforms (S900 and S1000+). As quoted by DJI:

> Each arm is designed with an 8° inversion and a 3° inclination, making the aircraft more stable when rolling and pitching, yet more flexible when rotating.

*Figure 6.14* DJI Spreading Wings range arms are designed with an 8° inversion (dihedral) angle and a 3° inclination (cant) angle. (By Frank Schwichtenberg – Own work, CC BY-SA 4.0, https://commons.wikimedia.org/w/index.php?curid=57790946).

As stated by DJI, these features are claimed to aid stability. However, they do so at the expense of lift. As always, there is a trade-off between one objective and another. At relatively small angles (< 10°), the penalty to be paid is fairly small.

The dihedral effect is used on many fixed-wing aircraft wings for roll stability and the cant angle has been successfully applied to many rotary-wing tail rotors for increased lift (30%) and CoG aftwards shift capability (see below).

There have been a small number of multi-rotor research projects that have taken the canting principle to the absolute limit. One of these was the work

*Figure 6.15* Sikorsky's Super Stallion (CH-53E) showing 20° canting of the tail rotor. Photograph by Jon 'ShakataGaNai' Davis.

of Dr William Crowther from the University of Manchester, who designed and built the 'Tumbleweed' hexrotor as an entry to the MoD Grand Challenge competition of 2008. In this military sponsored competition and trial, a number of original designs of UAV and UGV were developed for an urban warfare setting at Copehill Down.[6]

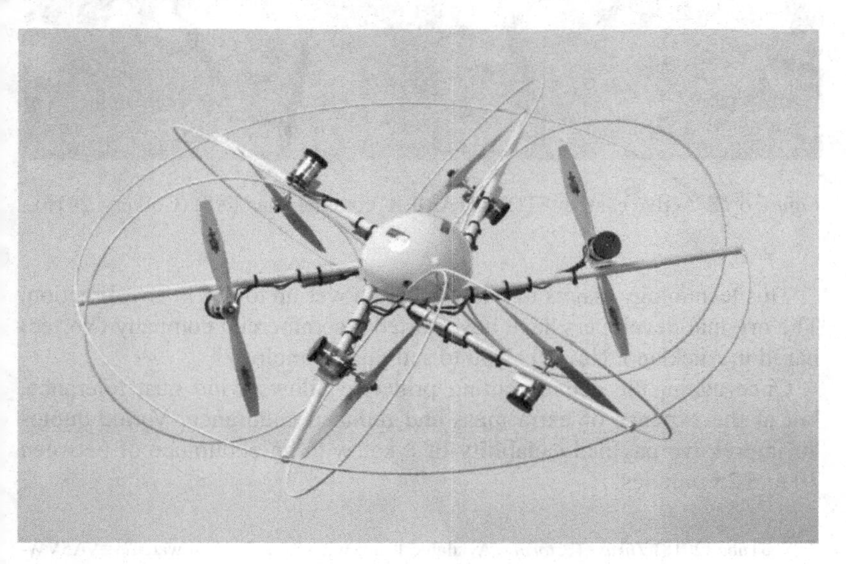

*Figure 6.16* Tumbleweed canted hexrotor (Crowther *et al.*, 2011).

6 Kuriositas. (2018) *Copehill Down*. Available from: www.kuriositas.com/2013/08/copehill-down-sleepy-english-village.html

## 6.5 Active thrust vectoring

Taking the static canting principle one step further, leads to active canting. This is sometimes referred to as thrust vectoring, and has been applied in full-scale military fighter aircraft for some time to aid rapid manoeuvrability.[7]

In the scale multi-rotor domain several researchers, including the author, have been inspired by the possibilities that this might allow. Of note, is the active canted co-axial quad (X8) originally developed by a postgraduate student, John Corkery from Churchill College, Cambridge in the UK (see below).[8]

*Figure 6.17* Active canted 'STORMTAMER' co-axial quad (X8) (Corkery, 2016).

This technology claims to allow angled hover up to 25° in any direction. The original developers have now started a commercial company (Vortec) based in Auckland, New Zealand to sell this technology.[9]

Once again, the active canting principle allows wind gust tolerance, but at the expense of extra mass and reduced endurance. Vortec quotes an impressive payload capability of 8 kg, with an endurance of between 10 and 25 minutes.

7 YouTube (2018) *Thrust Vectoring*. Available from: www.youtube.com/watch?v=VA5VW-qapzs

8 Corkery, J. (2016) Actively Canted Co-Axial Quad. Available from: www.chu.cam.ac.uk/news/2016/may/26/churchill-student-thrust-vector-drone/

9 Vortec (2017) Actively Canted Co-Axial Quad. Available from: https://vortec.nz/

# 7 Conclusion

Over the last 120 years, many developments in manned aircraft have taken place; however, it has really only been in the last two decades that technological developments in sensors and microprocessors have enabled small unmanned aircraft to become a reality. The use of such systems in the civilian domain is now a reality and this is causing much consternation amongst policy makers, governments and citizens as to what rules to put in place to enable this new industry to flourish, whilst protecting people's privacy and safety. With an open skies policy towards unmanned aircraft in the US in 2015 and the UK in 2016, this is now a pressing issue which has both an economic and societal impact.

## 7.1 Economic and societal impact

The widespread availability and use of low-cost small multi-rotor UA, has enabled researchers and citizen scientists to explore the three dimensional world in ways that only a few decades ago would have seemed impossible. Experiments in aerial photography, mapping and mineral exploration, environmental monitoring and disaster recovery from earthquakes, Tsunami, landslides and nuclear accidents, have all become a reality thanks to these type of systems. In developing the next generation of unmanned systems, such as the tethered small multi-rotor capable of persistent surveillance, developers will need to convince governmental agencies, like the Civil Aviation Authority, as well as the average citizen, of the benefits, safety and security of such systems, when used appropriately. A recent study has concluded that in the first 3 years of integration more than 70,000 jobs would be created in the United States with an economic impact of more than US$ 13.6bn. The benefit to the UK by 2030 has been projected to be a creation of 628,000 jobs with an economic impact of £ 42bn.[1]

1 PwC (2018) *Drone Report*. Available from: www.pwc.co.uk/press-room/press-releases/pwc-uk-drones-report.html

## 7.2 Future directions, opportunities and threats

As discussed throughout this book, great strides in aerial development have been achieved, by and large, all in the last 20 years. The pace of innovation and the opening up of the skies to entrepreneurial activity in the form of new delivery models, personal transportation and novel services, bodes well for the future city concept.

However, the old challenges still exist. These can be categorised as:

- Societal acceptance – Good use cases, need, awareness, insurance.
- Privacy issues – Big brother, security, filming, paparazzi.
- Safety – Terrorism, mid-air collision (MAC), sense & avoid.
- Reliability – Failure rate, GPS blocking, command and control (ATM).
- Cost/benefit ratio – Premium delivery, training, capability, range.
- Environmental resilience – Rain, wind, temperature, humidity.
- Legislation issues – CAA permissions, pilot licence, geo tagging, drone registration.

At the time of writing, many of the technical barriers of entry to the marketplace have been overcome. However, developers, manufacturers and operators are still unsure of where the legislation and public opinion will lead. It is clear from recent conversations that I have had with the public, that there is a greater understanding of the technology and the benefits to society of drone technology. Legislative bodies, such as the CAA in the UK have been at the forefront of developing drone regulation, but have been slow to prepare for the increasing interest and demand for commercial licences. The cost of a Permission for Commercial Operations (PfCO) recently increased dramatically to £ 173 for the first year, reducing to £ 130 at renewal. For special or non-standard permission this can cost £ 1211.

Despite the increases in charges, there continues to be a growing interest in commercial drone operations, with 4207 current Small Unmanned Aircraft (< 20 kg) permission holders registered with the UK CAA as of June 2018. Perhaps this is not surprising given the cost of a manned helicopter (£ 1–1.5k/hr).

Recent initiatives, such as NESTA's Flying High Challenge,[2] have focussed attention on flying drones within city centres. Five cities were selected for the phase 1 pilot study, which took place over four months from Feb–June 2018 and these were Southampton, London, Bradford, Preston and the West Midlands. Two of these cities focussed on delivery of medical products by drone.

2 NESTA (2018) *Flying High Challenge*. Available from: http://flyinghighchallenge.org/

In conclusion, the drone market is both buoyant and innovative with new use cases emerging on a daily basis. As we develop systems which are more capable, in terms of payload capacity, range (endurance) and efficiency we will need engineers with the skills and knowledge to be able to build and service these systems. As can be seen from the figure below, the best Li-Po powered drone can almost compete with the best Hydrogen Fuel Cell powered drone. However, the Hybrid Fuel-Battery powered drone has the edge, at the moment, in terms of performance. I hope that this book provides a good foundation for all such novel systems to be developed in the future.

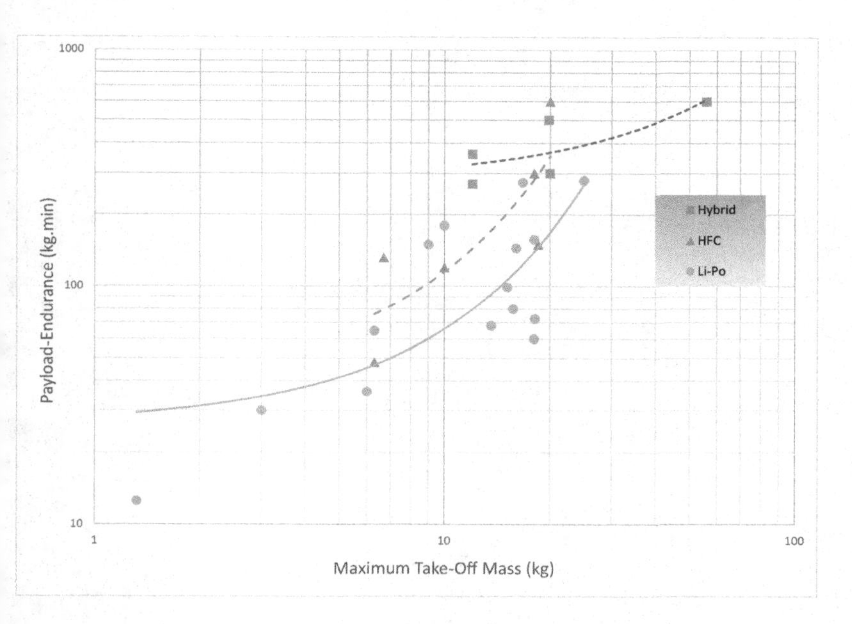


*Figure 7.1* Payload-Endurance vs MTOM for a range of heavy-lift multi-rotors.

# Li-Po batteries ready reckoner (6S)

*Power Density reckoner*

*Table showing battery mass (kg) for capacity (mAh) against desired power density (Wh/kg).*

*Nominal battery voltage* 22.2 *V*

| | *Power Density Wh/kg* | | | | | | | | | | |
|---|---|---|---|---|---|---|---|---|---|---|---|
| *mAh* | *150* | *155* | *160* | *165* | *170* | *175* | *180* | *185* | *190* | *195* | *200* |
| **5000** | 0.74 | 0.72 | 0.69 | 0.67 | 0.65 | 0.63 | 0.62 | 0.60 | 0.58 | 0.57 | 0.56 |
| **6000** | 0.89 | 0.86 | 0.83 | 0.81 | 0.78 | 0.76 | 0.74 | 0.72 | 0.70 | 0.68 | 0.67 |
| **7000** | 1.04 | 1.00 | 0.97 | 0.94 | 0.91 | 0.89 | 0.86 | 0.84 | 0.82 | 0.80 | 0.78 |
| **8000** | 1.18 | 1.15 | 1.11 | 1.08 | 1.04 | 1.01 | 0.99 | 0.96 | 0.93 | 0.91 | 0.89 |
| **9000** | 1.33 | 1.29 | 1.25 | 1.21 | 1.18 | 1.14 | 1.11 | 1.08 | 1.05 | 1.02 | 1.00 |
| **10000** | 1.48 | 1.43 | 1.39 | 1.35 | 1.31 | 1.27 | 1.23 | 1.20 | 1.17 | 1.14 | 1.11 |
| **11000** | 1.63 | 1.58 | 1.53 | 1.48 | 1.44 | 1.40 | 1.36 | 1.32 | 1.29 | 1.25 | 1.22 |
| **12000** | 1.78 | 1.72 | 1.67 | 1.61 | 1.57 | 1.52 | 1.48 | 1.44 | 1.40 | 1.37 | 1.33 |
| **13000** | 1.92 | 1.86 | 1.80 | 1.75 | 1.70 | 1.65 | 1.60 | 1.56 | 1.52 | 1.48 | 1.44 |
| **14000** | 2.07 | 2.01 | 1.94 | 1.88 | 1.83 | 1.78 | 1.73 | 1.68 | 1.64 | 1.59 | 1.55 |
| **15000** | 2.22 | 2.15 | 2.08 | 2.02 | 1.96 | 1.90 | 1.85 | 1.80 | 1.75 | 1.71 | 1.67 |
| **16000** | 2.37 | 2.29 | 2.22 | 2.15 | 2.09 | 2.03 | 1.97 | 1.92 | 1.87 | 1.82 | 1.78 |
| **17000** | 2.52 | 2.43 | 2.36 | 2.29 | 2.22 | 2.16 | 2.10 | 2.04 | 1.99 | 1.94 | 1.89 |
| **18000** | 2.66 | 2.58 | 2.50 | 2.42 | 2.35 | 2.28 | 2.22 | 2.16 | 2.10 | 2.05 | 2.00 |
| **19000** | 2.81 | 2.72 | 2.64 | 2.56 | 2.48 | 2.41 | 2.34 | 2.28 | 2.22 | 2.16 | 2.11 |
| **20000** | 2.96 | 2.86 | 2.78 | 2.69 | 2.61 | 2.54 | 2.47 | 2.40 | 2.34 | 2.28 | 2.22 |
| **21000** | 3.11 | 3.01 | 2.91 | 2.83 | 2.74 | 2.66 | 2.59 | 2.52 | 2.45 | 2.39 | 2.33 |
| **22000** | 3.26 | 3.15 | 3.05 | 2.96 | 2.87 | 2.79 | 2.71 | 2.64 | 2.57 | 2.50 | 2.44 |
| **23000** | 3.40 | 3.29 | 3.19 | 3.09 | 3.00 | 2.92 | 2.84 | 2.76 | 2.69 | 2.62 | 2.55 |
| **24000** | 3.55 | 3.44 | 3.33 | 3.23 | 3.13 | 3.04 | 2.96 | 2.88 | 2.80 | 2.73 | 2.66 |

# Appendix A2

## A range of 6S Li-Po batteries with high specific energy (Wh/kg) (June 2018)

| *6S Li-Po batteries >= 8500 mAh* | | | | | | *Rate $ to £* | | *Nominal voltage (V)* | | |
|---|---|---|---|---|---|---|---|---|---|---|
| | | | | | | *0.746* | | *22.2* | | |
| *Supplier* | *Manufacturer* | *Description* | *Mass (g)* | *Capacity (mAh)* | *Discharge (C)* | *$* | *£* | *Specific Energy (Wh/kg)* | *mAh/£p* | *Link* |
| MaxAmps | MaxAmps | Li-Po 23,000 6S 22.2 V Battery Pack | 2478 | 23000 | 25 | 699.99 | 522.19 | 206.05 | 0.44 | www.maxamps.com/lipo-23000-6s-22-2v-battery-pack |
| MaxAmps | MaxAmps | Li-Po 17,000 6S 22.2 V Battery Pack | 1928 | 17000 | 15 | 549.99 | 410.29 | 195.75 | 0.41 | www.maxamps.com/lipo-17000-6s-22-2v-battery-pack |
| Heliguy | Gens Tattu | GENS ACE Tattu 22,000 mAh 6S | 2509 | 22000 | 25 | | 359.99 | 194.66 | 0.61 | www.heliguy.com/gens-ace-tattu-22000mah-6s-p125 |
| MaxAmps | MaxAmps | Li-Po 11,000 6S 22.2 V Battery Pack | 1270 | 11000 | 40 | 399.99 | 298.39 | 192.28 | 0.37 | www.maxamps.com/lipo-11000-6s-22-2v-battery-pack |
| MaxAmps | MaxAmps | Li-Po 16,000 6S 22.2 V Battery Pack | 1860 | 16000 | 20 | 549.99 | 410.29 | 190.97 | 0.39 | www.maxamps.com/lipo-16000-6s-22-2v-battery-pack |
| Overlander | Overlander | 8500 mAh 6S2P 22.2 V 20C Li-Po Battery – Overlander Supersport XL | 1020 | 8500 | 20 | | 123.48 | 185.00 | 0.69 | wheelspinmodels.co.uk/i/262077/ |

| | | | | | | | | | | |
|---|---|---|---|---|---|---|---|---|---|---|
| Heliguy | Gens Tattu | GENS ACE Tattu 16000 mAh 6S | 1932 | 16000 | 15 | 325.79 | 243.04 | 183.85 | 0.66 | www.genstattu.com/tattu-16000mah-15c-6s1p-lipo-battery-pack-with-as150-xt150-plug.html |
| Gens Tattu | Gens Tattu | Tattu 16,000 mAh 15C 6S1P Li-Po Battery Pack with EC5 plug | 1932 | 16000 | 15 | 325.79 | 243.04 | 183.85 | 0.66 | www.genstattu.com/tattu-16000mah-15c-6s1p-lipo-battery-pack-with-ec5-plug.html |
| Overlander | Overlander | 22,000 mAh 6S 22.2 V 20C Li-Po Battery – Overlander Supersport XL | 2695 | 22000 | 20 | | 339.98 | 181.22 | 0.65 | www.overlander.co.uk/batteries/lipo-batteries/power-packs/22000mah-6s-22-2v-30c-lipo-battery-overlander-supersportxl.html |
| Overlander | Overlander | 20,000 mAh 6S2P 22.2 V 20C Li-Po Battery – Overlander Supersport XL | 2460 | 20000 | 20 | | 349.99 | 180.49 | 0.57 | www.overlander.co.uk/batteries/lipo-batteries/power-packs/20000mah-6s2p-22-2v-20c-lipo-battery-overlander-supersportxl.html |
| Overlander | Overlander | 10,000 mAh 6S 22.2 V 20C Li-Po Battery – Overlander Supersport XL | 1255 | 10000 | 20 | | 169.99 | 176.89 | 0.59 | www.overlander.co.uk/batteries/lipo-batteries/power-packs/10000mah-6s-22-2v-30c-lipo-battery-overlander-supersportxl.html |
| MaxAmps | MaxAmps | Li-Po 10,900 6S 22.2 V Battery Pack | 1371 | 10900 | 120 | 429.99 | 320.77 | 176.50 | 0.34 | www.maxamps.com/lipo-10900-6s-22-2v-battery-pack |
| Hobby King | MultiStar | Multistar High Capacity 20,000 mAh 6S 10C Multi-Rotor Li-Po Pack | 2519 | 20000 | 10 | | 112.47 | 176.26 | 1.78 | https://hobbyking.com/en_us/multistar-high-capacity-6s-20000mah-multi-rotor-lipo-pack.html |

(*Continued*)

(Continued)

| *6S Li-Po batteries >= 8500 mAh* | | | | | | *Rate $ to £* | | *Nominal voltage (V)* | | |
|---|---|---|---|---|---|---|---|---|---|---|
| | | | | | | *0.746* | | *22.2* | | |
| *Supplier* | *Manufacturer* | *Description* | *Mass (g)* | *Capacity (mAh)* | *Discharge (C)* | *$* | *£* | *Specific Energy (Wh/kg)* | *mAh/£p* | *Link* |
| Hobby King | MultiStar | Multistar High Capacity 16,000 mAh 6S 10C Multi-Rotor Li-Po Pack | 2044 | 16000 | 10 | | 92.38 | 173.78 | 1.73 | https://hobbyking.com/en_us/multistar-high-capacity-6s-16000mah-multi-rotor-lipo-pack.html |
| Heliguy | Gens Tattu | GENS ACE Tattu 10,000 mAh 6S 15C | 1300 | 10000 | 15 | | 148.00 | 170.77 | 0.68 | www.heliguy.com/gens-ace-tattu-10000mah-6s-15c-p238 |
| Overlander | Overlander | 12,500 mAh 6S2P 22.2 V 35C Li-Po Battery – Overlander Supersport XL | 1640 | 12500 | 35 | | 259.99 | 169.21 | 0.48 | www.overlander.co.uk/batteries/lipo-batteries/power-packs/lipo-batteries-12500mah-6s2p-22-2v-35c-supersportxl.html |
| Heliguy | Gens Tattu | GENS ACE Tattu 12,000 mAh 6S 15C | 1577 | 12000 | 15 | | 188.00 | 168.93 | 0.64 | www.heliguy.com/gens-ace-tattu-12000mah-6s-15c-p208 |
| Heliguy | Gens Tattu | Tattu PLUS 10,000 mAh 6S 25C BATTERIES | 1319 | 10000 | 20 | 196.31 | 146.45 | 168.31 | 0.68 | www.genstattu.com/tattu-10000mah-22-2v-20c-6s1p-lipo-battery-pack-with-xt60-plug.html |
| Hobby King | MultiStar | MultiStar High Capacity 12,000 mAh 6S 10C Multi-Rotor Li-Po Pack | 1602 | 12000 | 10 | 110 | 79.61 | 166.29 | 1.51 | https://hobbyking.com/en_us/multistar-high-capacity-6s-12000mah-multi-rotor-lipo-pack.html?wrh_pdp=1 |
| Gens Tattu | Gens Tattu | Tattu 12,000 mAh 6S1P 15C Li-Po Battery Pack with XT90 plug | 1620 | 12000 | 15 | 156.38 | 116.66 | 164.44 | 1.03 | www.genstattu.com/tattu-12000mah-6s1p-15c-lipo-battery-pack-with-xt90-plug.html |

# Appendix B

## Guidance on the safe use, handling, storage and disposal of Lithium Polymer batteries

It is strongly recommended that before charging and/or using a Lithium Polymer battery, you should read and follow this guidance. Failure to do so may result in a fire, leading potentially to personal injury and damage to, or loss of, assets.

### 1 General safety instructions and warnings

1.1 When charging, only use a Lithium Polymer battery charger. Do not use a Nickel Metal Hydride or Nickel Cadmium charger. Failure to do so may cause a fire.

1.2 Never charge batteries unattended. When charging Lithium Polymer batteries you should always observe and monitor the charging process so that you can react to potential problems that may occur.

1.3 Some Lithium Polymer chargers on the market may have technical deficiencies that may cause the Lithium Polymer batteries to be charged incorrectly or at an improper rate. It is the user's responsibility to ensure that the charger works properly and is PAT tested. Always monitor the charging process to ensure batteries are being charged properly. Failure to do so may result in a fire.

1.4 If at any time you witness a battery starting to balloon or swell up, discontinue the charging process immediately, disconnect the battery, remove it to a safe area and leave it, preferably under observation, for approximately 15 minutes. This swelling may cause the battery to leak, and the reaction with air may cause the chemicals to ignite, resulting in a fire.

1.5 Since delayed chemical reaction can occur, it is best to continue to observe the battery as a precaution. This should only be undertaken in a safe area outside of any building or vehicle and away from any combustible material.

1.6 Wire lead shorts can cause fires. If you accidentally short the wires, the battery must be placed in a safe area for observation for approximately 15 minutes. Additionally, if a short occurs and contact is made with metal (such as a ring on a finger), severe injuries may occur due to the conductibility of electric current.

1.7. If for any reason you need to cut the terminal wires, it will be necessary to cut each wire separately, ensuring the wires do not touch each other causing a short and potentially causing a fire.
1.8. Always ensure that appropriate fire-fighting equipment[1] is available and close at hand in the battery charging area or in the place where damaged or swollen batteries are placed for observation.
1.9. Lithium Polymer batteries are volatile and can react to extreme temperatures. Never store, transport or charge battery packs inside your car in extreme temperatures, since these can cause a battery to ignite.

## 2 The charging process

2.1 Never leave batteries charging unattended.
2.2 Charge in an isolated area, appropriately signed and away from other flammable materials.
2.3 Let the battery cool down to ambient temperature before charging.
2.4 Do not charge battery packs in series. Charge each battery pack individually. Failure to do so may result in incorrect battery recognition and charging functions. Overcharging may occur and a fire may result.
2.5 When selecting the cell count or voltage for charging purposes, select the cell count and voltage as it appears on the battery label. As a safety precaution, please confirm the information printed on the battery is correct.

  (a) Example: The label on a 2-Cell battery pack in series will read – 'Charge as 2-Cell (7.4 V), or may cause fire' – You must select 2-Cell for charging.
  (b) Example: The label on a 3-Cell battery pack in series will read – 'Charge as 3-Cell (11.1 V), or may cause fire' – You must select 3-Cell for charging.

2.6 Selecting a cell count other than the one printed on the battery (always confirm label is correct), can cause a fire.
2.7 You must check the pack voltage before charging. Do not attempt to charge any pack if open voltage per cell is less than 3.3 V.

**Examples:**

(a) Do not charge a 2-Cell pack if below 6.6 V.
(b) Do not charge a 3-Cell pack if below 9.9 V.

1 Safelincs (2017). Available from: www.safelincs.co.uk/lith-ex-fire-extinguisher/

2.8 You must select the charge rate current that does not exceed the recommended 'C' rating of the battery – usually written on the side of the battery. A higher setting may cause a fire.

**First Discharge:** Keep to 6-minute sessions with 15-minute breaks.

## 3 Storage and transportation

3.1 Store batteries at room temperature between 40 and 80 degrees F, (4 and 27 degrees C) for best results.
3.2 Do not expose battery packs to direct sunlight (heat) for extended periods.
3.3 When transporting or temporarily storing in a vehicle, the temperature range should be greater than 20 degrees F, (−7 degrees C) but no more than 150 degrees F, (66 degrees C).
3.4 Storing batteries at temperatures greater than 170 degrees F, (77 degrees C) for extended periods of time (for more than 2 hours), may cause damage to the battery and possibly cause a fire.

## 4 Caring for your Lithium Polymer battery

4.1 Charge batteries using a good quality Lithium Polymer charger. A poor quality charger can be dangerous.
4.2 Make sure the voltage and current are set correctly, as failure to do so can cause a fire.
4.3 Please check cell voltage after the first charge.

**Examples**

- 1-Cell: 4.2 V (4.15–4.22 V)
- 2-Cell: 8.4 V (8.30–8.44 V)
- 3-Cell: 12.6 V (12.45–12.66 V)
- 4-Cell: 16.8 V (16.60–16.88 V)
- 5-Cell: 21.0 V (20.75–21.1 V)
- 6-Cell: 25.2 V (24.90–25.32 V)

4.4 Do not discharge the battery to a level below 3 V per cell under load. Deep discharge below 3 V per cell can deteriorate battery performance.
4.5 Use caution when handling Lithium Polymer batteries to avoid puncturing the cells, as this may cause a fire.

**Operating Temperature**

- **Charge:**
- 32–113 degrees F (0–45 degrees C)

- Let the battery cool down to an ambient temperature before charging.
- **Discharge:**
- 32–140 degrees F (0–60 degrees C)
- During discharge and handling of batteries, do not exceed 160 degrees F (71 degrees C).
- **Battery Life**
- Batteries that lose 20% of their capacity must be removed from service and disposed of properly.
- Discharge the battery to 3 V per Cell, making sure output wires are insulated, then wrap the battery in a bag for disposal.

## 5 Safe disposal of Lithium Polymer batteries

Lithium Polymer batteries are used in a wide variety of both scientific equipment and personal electrical equipment. There are several types of Lithium batteries, but they are all high energy power sources, and all are potentially hazardous.

For safety reasons, it is best that Lithium Polymer cells be fully discharged before disposal.

However, if physically damaged, it is NOT recommended to discharge Lithium Polymer cells before disposal, (see below for details). The batteries must also be cool before proceeding with disposal instructions. To dispose of Lithium Polymer cells and packs the instructions below should be followed.

5.1 If any Lithium Polymer cell in the pack has been physically damaged, resulting in a swollen cell or a split or tear in a cell's foil covering, do NOT discharge the battery. Go to Step 5.5.

5.2 Place the Lithium Polymer battery in a fireproof container or bucket of sand.

5.3 Connect the battery to a Lithium Polymer discharger. Set the discharge cut-off voltage to the lowest possible value. Set the discharge current to a C/10 value, with 'C' being the capacity rating of the pack. For example, the '10C' rating for a 1200 mAh battery is 12 A, and that battery's C/10 current value is (12 A / 10) 1.2 A or 1200 mA. Or, a simple resistive type of discharge load can be used, such as a power resistor or set of light bulbs as long as the discharge current doesn't exceed the C/10 value and cause an overheating condition.

For Lithium Polymer packs rated at 7.4 and 11.1 V, connect a 150 ohm resistor with a power rating of 2 W to the pack's positive and negative terminals to safely discharge the battery.

It is also possible to discharge the battery by connecting it to an ESC/motor system and allowing the motor to run indefinitely until no power remains to further cause the system to function.

5.4 Submerge the battery into a bucket or tub of salt water. This container should have a lid, but it does not need to be air-tight. Prepare a bucket or tub containing 3–5 gallons of cold water, and mix in 1/2 cup of salt per gallon of water. Drop the battery into the salt water. Allow the battery to remain in the tub of salt water for at least 2 weeks.

5.5 Remove the Lithium Polymer battery from the salt water and dispose of as hazardous waste. Written confirmation that the batteries have been fully discharged and soaked for the appropriate length of time, should accompany the batteries to the hazardous waste storage facility.

# T-Motor BLDC motor selection tool

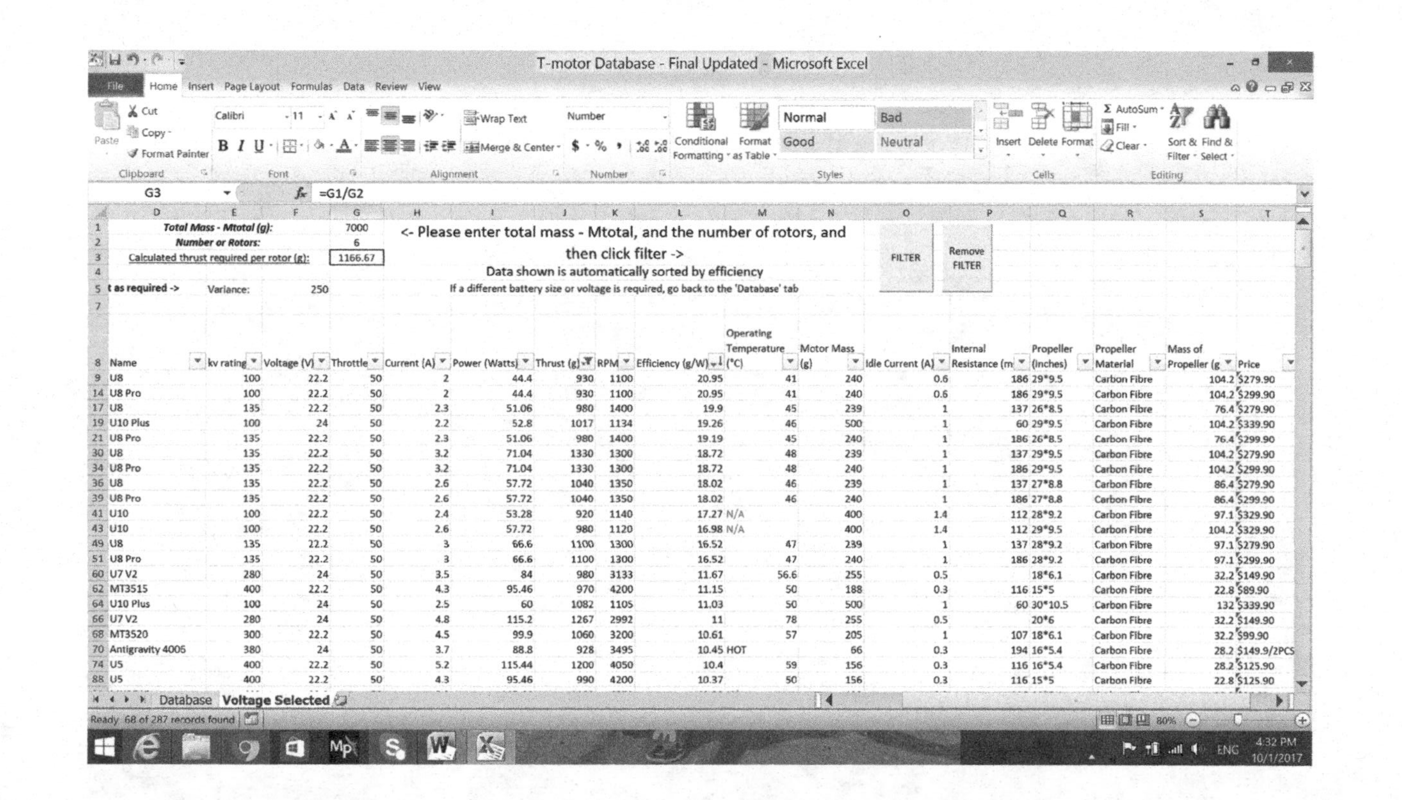


| Row | Name | kv rating | Voltage (V) | Throttle | Current (A) | Power (Watts) | Thrust (g) | RPM | Efficiency (g/W) | Operating Temperature (°C) | Motor Mass (g) | Idle Current (A) | Internal Resistance (m | Propeller (inches) | Propeller Material | Mass of Propeller (g | Price |
|---|---|---|---|---|---|---|---|---|---|---|---|---|---|---|---|---|---|
| 9 | U8 | 100 | 22.2 | 50 | 2 | 44.4 | 930 | 1100 | 20.95 | 41 | 240 | 0.6 | 186 | 29*9.5 | Carbon Fibre | 104.2 | $279.90 |
| 14 | U8 Pro | 100 | 22.2 | 50 | 2 | 44.4 | 930 | 1100 | 20.95 | 41 | 240 | 0.6 | 186 | 29*9.5 | Carbon Fibre | 104.2 | $299.90 |
| 17 | U8 | 135 | 22.2 | 50 | 2.3 | 51.06 | 980 | 1400 | 19.9 | 45 | 239 | 1 | 137 | 26*8.5 | Carbon Fibre | 76.4 | $279.90 |
| 19 | U10 Plus | 100 | 24 | 50 | 2.2 | 52.8 | 1017 | 1134 | 19.26 | 46 | 500 | 1 | 60 | 29*9.5 | Carbon Fibre | 104.2 | $339.90 |
| 21 | U8 Pro | 135 | 22.2 | 50 | 2.3 | 51.06 | 980 | 1400 | 19.19 | 45 | 240 | 1 | 186 | 26*8.5 | Carbon Fibre | 76.4 | $299.90 |
| 30 | U8 | 135 | 22.2 | 50 | 3.2 | 71.04 | 1330 | 1300 | 18.72 | 48 | 239 | 1 | 137 | 29*9.5 | Carbon Fibre | 104.2 | $279.90 |
| 34 | U8 Pro | 135 | 22.2 | 50 | 3.2 | 71.04 | 1330 | 1300 | 18.72 | 48 | 240 | 1 | 186 | 29*9.5 | Carbon Fibre | 104.2 | $299.90 |
| 36 | U8 | 135 | 22.2 | 50 | 2.6 | 57.72 | 1040 | 1350 | 18.02 | 46 | 239 | 1 | 137 | 27*8.8 | Carbon Fibre | 86.4 | $279.90 |
| 39 | U8 Pro | 135 | 22.2 | 50 | 2.6 | 57.72 | 1040 | 1350 | 18.02 | 46 | 240 | 1 | 186 | 27*8.8 | Carbon Fibre | 86.4 | $299.90 |
| 41 | U10 | 100 | 22.2 | 50 | 2.4 | 53.28 | 920 | 1140 | 17.27 | N/A | 400 | 1.4 | 112 | 28*9.2 | Carbon Fibre | 97.1 | $329.90 |
| 43 | U10 | 100 | 22.2 | 50 | 2.6 | 57.72 | 980 | 1120 | 16.98 | N/A | 400 | 1.4 | 112 | 29*9.5 | Carbon Fibre | 104.2 | $329.90 |
| 49 | U8 | 135 | 22.2 | 50 | 3 | 66.6 | 1100 | 1300 | 16.52 | 47 | 239 | 1 | 137 | 28*9.2 | Carbon Fibre | 97.1 | $279.90 |
| 51 | U8 Pro | 135 | 22.2 | 50 | 3 | 66.6 | 1100 | 1300 | 16.52 | 47 | 240 | 1 | 186 | 28*9.2 | Carbon Fibre | 97.1 | $299.90 |
| 60 | U7 V2 | 280 | 24 | 50 | 3.5 | 84 | 980 | 3133 | 11.67 | 56.6 | 255 | 0.5 |  | 18*6.1 | Carbon Fibre | 32.2 | $149.90 |
| 62 | MT3515 | 400 | 22.2 | 50 | 4.3 | 95.46 | 970 | 4200 | 11.15 | 50 | 188 | 0.3 | 116 | 15*5 | Carbon Fibre | 22.8 | $89.90 |
| 64 | U10 Plus | 100 | 24 | 50 | 2.5 | 60 | 1082 | 1105 | 11.03 | 50 | 500 | 1 | 60 | 30*10.5 | Carbon Fibre | 132 | $339.90 |
| 66 | U7 V2 | 280 | 24 | 50 | 4.8 | 115.2 | 1267 | 2992 | 11 | 78 | 255 | 0.5 |  | 20*6 | Carbon Fibre | 32.2 | $149.90 |
| 68 | MT3520 | 300 | 22.2 | 50 | 4.5 | 99.9 | 1060 | 3200 | 10.61 | 57 | 205 | 1 | 107 | 18*6.1 | Carbon Fibre | 32.2 | $99.90 |
| 70 | Antigravity 4006 | 380 | 24 | 50 | 3.7 | 88.8 | 928 | 3495 | 10.45 | HOT | 66 | 0.3 | 194 | 16*5.4 | Carbon Fibre | 28.2 | $149.9/2PCS |
| 74 | U5 | 400 | 22.2 | 50 | 5.2 | 115.44 | 1200 | 4050 | 10.4 | 59 | 156 | 0.3 | 116 | 16*5.4 | Carbon Fibre | 28.2 | $125.90 |
| 88 | U5 | 400 | 22.2 | 50 | 4.3 | 95.46 | 990 | 4200 | 10.37 | 50 | 156 | 0.3 | 116 | 15*5 | Carbon Fibre | 22.8 | $125.90 |

# Overall system selection spreadsheet

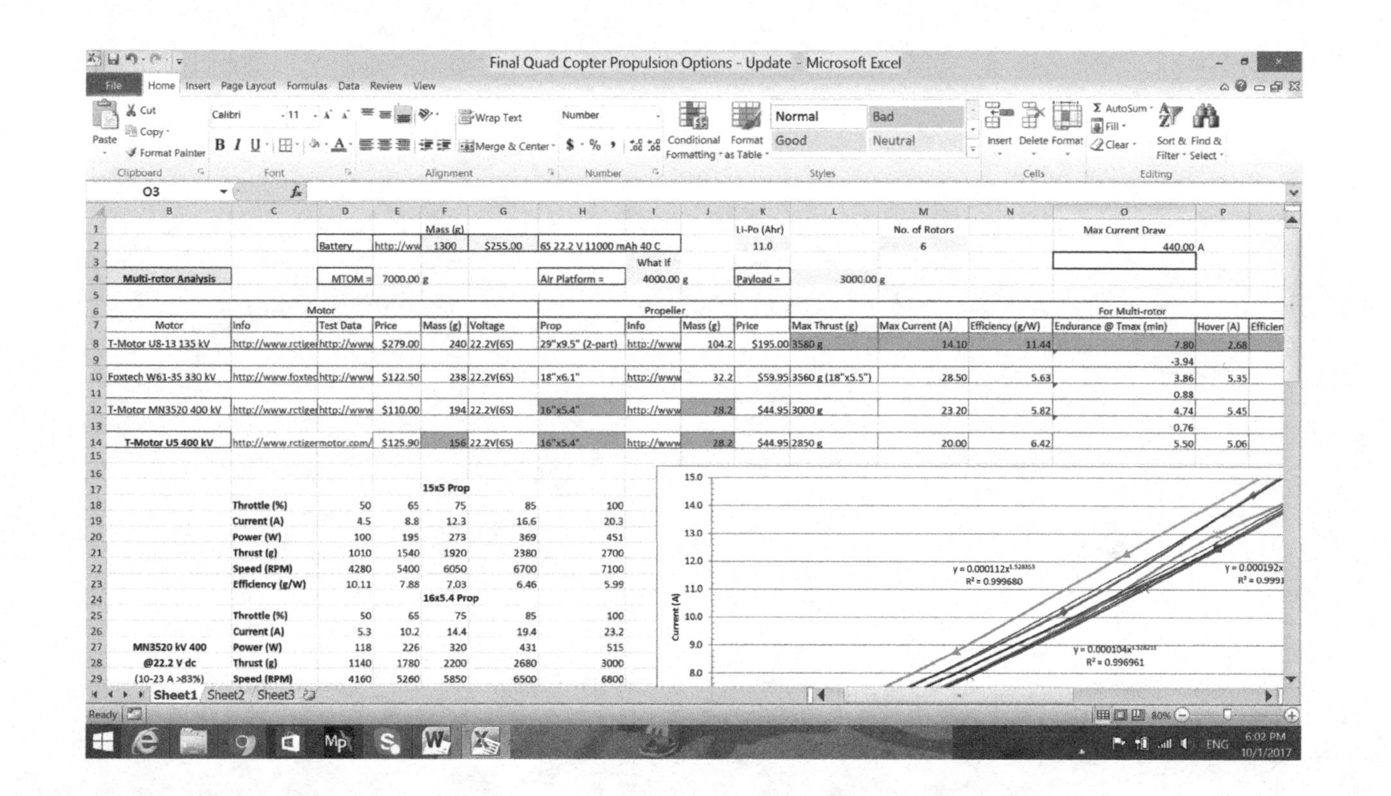

Final Quad Copter Propulsion Options - Update - Microsoft Excel

| | | | | Mass (g) | | | | | Li-Po (Ahr) | | No. of Rotors | | Max Current Draw |
|---|---|---|---|---|---|---|---|---|---|---|---|---|---|
| | | Battery | http://ww | 1300 | $255.00 | 6S 22.2 V 11000 mAh 40 C | | | 11.0 | | 6 | | 440.00 A |
| | | | | | | | What If | | | | | | |
| Multi-rotor Analysis | | MTOM = | 7000.00 g | | | Air Platform = | 4000.00 g | | Payload = | 3000.00 g | | | |

| Motor | | | | | | Propeller | | | | For Multi-rotor | | | | | |
|---|---|---|---|---|---|---|---|---|---|---|---|---|---|---|---|
| Motor | Info | Test Data | Price | Mass (g) | Voltage | Prop | Info | Mass (g) | Price | Max Thrust (g) | Max Current (A) | Efficiency (g/W) | Endurance @ Tmax (min) | Hover (A) | Efficien |
| T-Motor U8-13 135 kV | http://www.rctiger | http://www | $279.00 | 240 | 22.2V(6S) | 29"x9.5" (2-part) | http://www | 104.2 | $195.00 | 3580 g | 14.10 | 11.44 | 7.80 | 2.68 | |
| | | | | | | | | | | | | | -3.94 | | |
| Foxtech W61-35 330 kV | http://www.foxtec | http://www | $122.50 | 238 | 22.2V(6S) | 18"x6.1" | http://www | 32.2 | $59.95 | 3560 g (18"x5.5") | 28.50 | 5.63 | 3.86 | 5.35 | |
| | | | | | | | | | | | | | 0.88 | | |
| T-Motor MN3520 400 kV | http://www.rctiger | http://www | $110.00 | 194 | 22.2V(6S) | 16"x5.4" | http://www | 28.2 | $44.95 | 3000 g | 23.20 | 5.82 | 4.74 | 5.45 | |
| | | | | | | | | | | | | | 0.76 | | |
| T-Motor U5 400 kV | http://www.rctigermotor.com/ | | $125.90 | 156 | 22.2V(6S) | 16"x5.4" | http://www | 28.2 | $44.95 | 2850 g | 20.00 | 6.42 | 5.50 | 5.06 | |

| | | | 15x5 Prop | | |
|---|---|---|---|---|---|
| | Throttle (%) | 50 | 65 | 75 | 85 | 100 |
| | Current (A) | 4.5 | 8.8 | 12.3 | 16.6 | 20.3 |
| | Power (W) | 100 | 195 | 273 | 369 | 451 |
| | Thrust (g) | 1010 | 1540 | 1920 | 2380 | 2700 |
| | Speed (RPM) | 4280 | 5400 | 6050 | 6700 | 7100 |
| | Efficiency (g/W) | 10.11 | 7.88 | 7.03 | 6.46 | 5.99 |
| | | | 16x5.4 Prop | | | |
| | Throttle (%) | 50 | 65 | 75 | 85 | 100 |
| | Current (A) | 5.3 | 10.2 | 14.4 | 19.4 | 23.2 |
| MN3520 kV 400 | Power (W) | 118 | 226 | 320 | 431 | 515 |
| @22.2 V dc | Thrust (g) | 1140 | 1780 | 2200 | 2680 | 3000 |
| (10-23 A >83%) | Speed (RPM) | 4160 | 5260 | 5850 | 6500 | 6800 |

# Appendix E
Examples of multi-rotor systems built at the ASL

**DARPA/SPARWAR UAVForge challenge 2012**

Team Name: **HALO (Overall Winner)**[1]
Max Take-Off Mass (kg) – w/out battery or payload: 1.65 kg
Number of motors: Six (Co-Axial) (Y6) (Detachable Arms)
Chosen Li-Po capacity (mAh) and mass (kg): (4S) 11,000 mAh and 0.85 kg (191.5 Wh/kg)
Payload: 0.25 kg (small cameras and transmitter)

1 http://archive.darpa.mil/uavforge/

Motor Model Number and URL:

RCTimer D5010-14 360 $K_v$ www.rctimer.com/?product-1080.html

Prop Size (Diameter x Pitch (inches)) and URL: 16″ × 5.5″ CF propellers www.rctimer.com/?product-841.html
Current draw per motor (A) @ hover (no payload, but with chosen Li-Po):

2.4 A theoretically, but we assume practically it probably won´t go under 3 A.

Equation of Motor Current (A)(Y-Axis) Vs Thrust (g)(X-Axis) (in the form of Current = $aX^b$): $y = 0.0032X^{1.1275}$
Overall one-off cost (£) (parts only): £ 3600 with VAT (20%) – DJI WK-M Flight Controller.
Endurance (min) for 0.25 kg payload and 4S 11,000 mAh MaxAmps Li-Po battery: 32 min

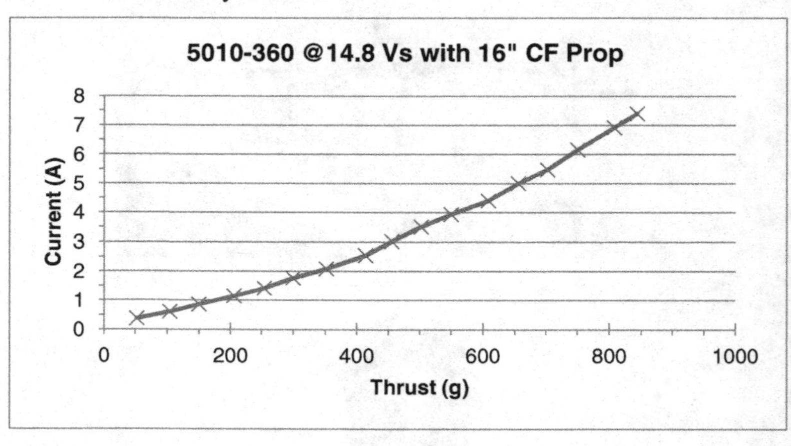

**IMechE UAS challenge entry 2016**

Team Name: **Daedalus (Overall 2nd Place)**
Max Take-Off Mass (kg) – w/out battery or payload: 2.934 kg
Number of motors: four (Detachable Arms)
Chosen Li-Po capacity (mAh) and mass (kg): (6S) 10,000 mAh, 1.204 kg (184.4 Wh/kg)
Payload: 2 kg
Motor Model Number and URL: Temporary Motors – Foxtech W61-35 KV330
www.goodluckbuy.com/foxtech-w61-35-kv330-brushless-motor-cw-ccw-combo-formulticopter.html
Prop Size (Diameter x Pitch (inches)) and URL: 18″ × 6.1″
www.hobbyking.com/hobbyking/store/__58197__Multistar_Timber_T_Style_Propeller_18x6_1_White_CW_CCW_2pcs_.html
Current draw per motor (A) @ hover (no payload, but with chosen Li-Po): 3.8 A
Equation of Motor Current (A)(Y-Axis) Vs Thrust (g)(X-Axis) (in the form of Current = $aX^b$):

$$(Y = 000001x^2 + 0.003111x - 0.181123)$$

Show a graph of this curve:
Overall one-off cost (£) (parts only): £ 978.97 with MN5212 motors – Pixhawk FC.

Endurance (min) for 2 kg payload and 12,000 mAh Multi-Star Li-Po battery: 18.15 min

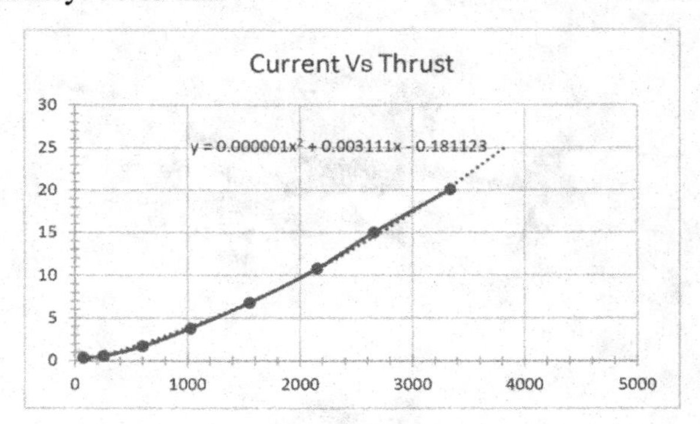

## IMechE UAS challenge entry 2016

Team Name: **Saffire (Overall 3rd Place)**
Max Take-Off Mass (kg) – w/out battery or payload: 3.25 kg
Number of motors: Six (Hexrotor) (three Inverted Rotors)
Chosen Li-Po capacity (mAh) and mass (kg): (6S) 9000 mAh and 1.25 kg (159.8 Wh/kg)
Payload: 2 kg
Motor Model Number and URL:

X5 (3525) 400 $K_v$ http://rctimer.com/product-1191.html
http://rctimer.com/product-1291.html

Prop Size (Diameter x Pitch (inches)) and URL: 15″ × 5.2″ foldable propellers
www.hobbyking.com/hobbyking/store/__62475__Multirotor_Folding_Propeller_15x5_2_Black_CW_CCW_4pcs_.html
Current draw per motor (A) @ hover (no payload, but with chosen Li-Po):

2.7 A theoretically, but we assume practically it probably won´t go under 4 A.

Equation of Motor Current (A)(Y-Axis) Vs Thrust (g)(X-Axis) (in the form of Current = aXb):

y = 0.000180X1.494881

Overall one-off cost (£) (parts only): £ 892.94 with VAT (20%) – Pixhawk FC.

Endurance (min) for 2 kg payload and 12,000 mAh Multi-Star Li-Po battery: 20.2 min

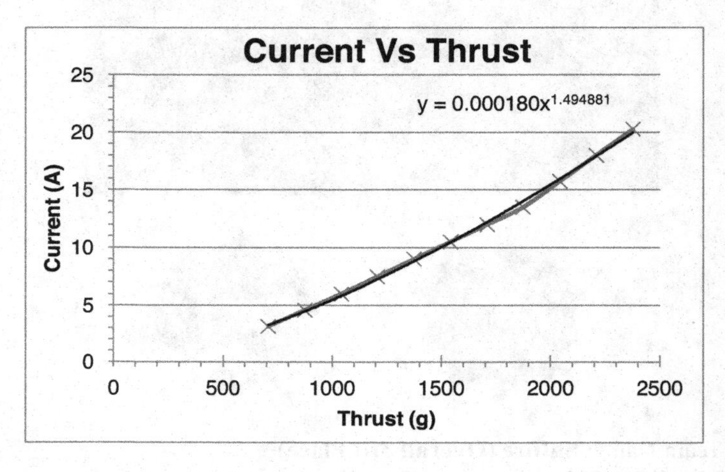

## IMechE UAS challenge entry 2016

Team Name: **Gryphon (Overall 4th Place)**
Max Take-Off Mass (kg) – w/out battery or payload: 3.0 kg
Number of motors: Eight (Co-Axial Quad) (Folding Arms/Legs)
Chosen Li-Po capacity (mAh) and mass (kg): (6S) 10,000 mAh and 1.22 kg (182 Wh/kg)
Payload: 2 kg
Motor Model Number and URL:

Tiger Motor MN4010 370 KV

Prop Size (Diameter x Pitch (inches)) and URL: 16″ × 5.5″ (Top) & 15″ × 6.5″ (Bottom)
www.RCTimer.com
Current draw per motor (A) @ hover (no payload, but with chosen Li-Po):

3 A theoretically, but we assume practically it probably won´t go under 5 A.

Equation of Motor Current (A)(Y-Axis) Vs Thrust (g)(X-Axis) (in the form of Current = $aX^b$):

$$y = 0.000180X^{1.494881}$$

Overall one-off cost (£) (parts only): £ 965 (incl. VAT @20%) – Pixhawk FC.

Endurance (min) for 2 kg payload and 10,000 mAh Multi-Star Li-Po battery: 14 min

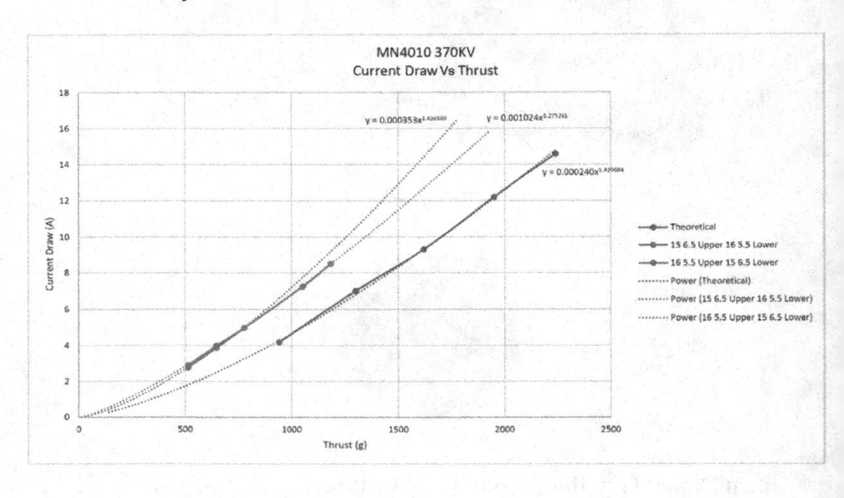

## IMechE UAS challenge entry 2017

Team Name: **Valkyrie (Overall 2nd Place)**
Max Take-Off Mass (kg) – w/out battery or payload: 2.891 kg
Number of motors: Six (Foldable Arms – Two Detachable)
Chosen Li-Po capacity (mAh) and mass (kg): Multistar (6S) 12,000 mAh and 1.328 kg (200.6 Wh/kg)
Payload: 2.5 kg
Motor Model Number and URL:

**Tiger T-Motor MN4014-9 400** $K_v$
www.quadcopters.co.uk/tiger-t-motor-mn4014-9-400kv-brushless-motor--3-left-in-stock-670-p.asp

Prop Size (Diameter x Pitch (inches)) and URL: **16″** × **5.5″** http://rctimer.com/product-923.html
Current draw per motor (A) @ hover (no payload, but with chosen Li-Po):

**1.93 A per motor (at a system mass = 4.219 kg)**

Equation of Motor Current (A)(Y-Axis) Vs Thrust (g)(X-Axis) (in the form of Current = $aX^b$):

(Example: $0.000112X^{1.528353}$)
$\mathbf{I_{motor} = 5e\text{-}5X^{1.6113}}$ **(X = thrust in grams)**

Overall one-off cost (£) (parts only): **£ 874.43** – Pixhawk FC.
Endurance (min) for 2, 2.5, 3 and 3.5 kg payload and 12,000 mAh Multistar Li-Po battery:

(State if this would exceed the 7 kg MTOM limit).

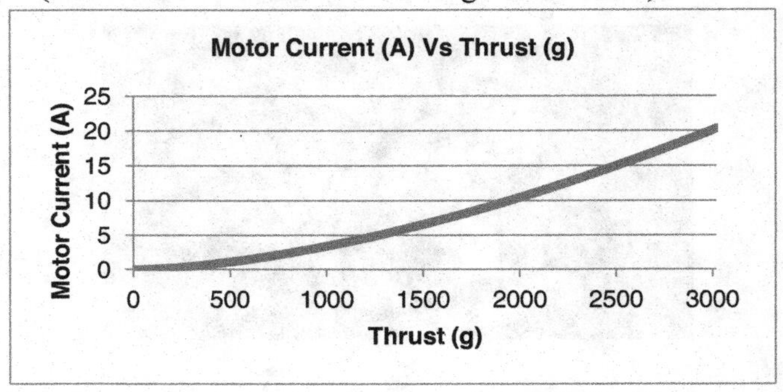


| *Payload (kg)* | *Endurance (to 80% Discharge of 12,000 mAh Multistar at Thrust/ Weight = 1) (min)* | *MTOW (kg)* | *Under/Over 7 kg Limit* |
|---|---|---|---|
| **2** | 26.6 | 6.22 | Under Limit |
| **2.5** | 23.5 | 6.72 | Under Limit |
| **3** | 20.9 | 7.22 | Over Limit (0.22 kg over) |
| **3.5** | 18.8 | 7.72 | Over Limit (0.72 kg over) |

**IMechE UAS challenge entry 2017**

Team Name: **Olympus (Overall 3rd Place)**
Max Take-Off Mass (kg) – without battery or payload: 2.8 kg
Number of motors: four
Chosen Li-Po capacity (mAh) and mass (kg): (6S) 12,000 mAh and 1.33 kg (200.3 Wh/kg)
Payload: 3 kg
Motor Model Number and URL:

T-Motor MN4014 330KV
www.rctigermotor.com/html/2013/Navigator_0910/40.html

Prop Size (Diameter x Pitch (inches)) and URL:

Tarot High Efficiency 18″ × 5.5″ CF Set TL2848 (Pair)
www.flyingtech.co.uk/frames-props-parts-accessories/tarot-1855-18x55-carbon-fibre-propeller-set-cwccw-tl2848

Current draw per motor (A) @ hover (no payload, but with chosen Li-Po): 3.6 A
(4.1 kg mass without payload, thrust per motor 1060 g, 45% throttle)
Equation of Motor Current (A)(Y-Axis) Vs Thrust (g)(X-Axis) (in the form of Current = $aX^b$):

(Example: $0.000112X^{1.528353}$)

Overall one-off cost (£) (parts only): £ 720 – Pixhawk FC.
Endurance (min) for 2, 2.5, 3 and 3.5 kg payload and 12,000 mAh Multi-Star Li-Po battery:

(State if this would exceed the 7 kg MTOM limit)

Theoretical Endurance using data extrapolated/interpolated from T-Motors datasheet:

Airframe mass: 2.8 kg
Battery mass: 1.328 kg

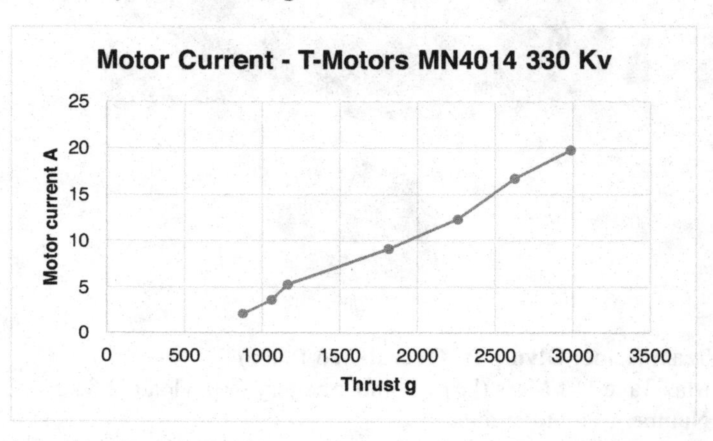


Endurance using 12,000 mAh battery

| *Payload (kg)* | *Theoretical Endurance (minutes)* | *Takeoff Mass (kg)* | *Exceeds 7 kg MTOM limit?* |
|---|---|---|---|
| 2 | 21 | 6.128 | No |
| 2.5 | 18.5 | 6.628 | No |
| 3 | 16.5 | 7.128 | Yes |
| 3.5 | 15 | 7.628 | Yes |

Airframe mass: 2.8 kg
Battery mass: 1.200 kg

Endurance using 10,000 mAh battery

| *Payload (kg)* | *Theoretical Endurance (minutes)* | *Take-off Mass (kg)* | *Exceeds 7 kg MTOM limit?* |
|---|---|---|---|
| 2 | 17 | 6.0 | No |
| 2.5 | 15 | 6.5 | No |
| 3 | 13.25 | 7.0 | No |
| 3.5 | 12 | 7.5 | Yes |

## IMechE UAS challenge entry 2018

Team Name: **Athena (Overall 2nd Place)**
Maximum Take-Off Mass (kg) – without battery or payload: 2.4 kg
Number of motors: Six (Y6) (Folding 'Crook' Arms)
Chosen Li-Po capacity (mAh) and mass (kg): (6S) 8500 mAh and 1.020 kg (185 Wh/kg)
Payload (kg): 3.5 kg
Motor Model Number and URL:

Tarot MT4008 330 kV (www.helipal.com/tarot-mt4008-high-power-brushless-motor330kv.html)

Prop Size (Diameter x Pitch (inches)) and URL:

- 17″ × 5.8″ T-Motor (http://store-en.tmotor.com/goods.php?id=383)
- 18″ × 6.1″ T-Motor (http://store-en.tmotor.com/goods.php?id=384)

Current draw per motor (A) @ hover (no payload, but with chosen Li-Po):

- Average current draw with no Payload: 11 A (3.6 kg MTOM)
- Average current draw with 3 kg Payload: 28 A (6.64 kg MTOM)

Equation of Motor Current (A)(Y-Axis) Vs Thrust (g)(X-Axis) (in the form of Current = $aX^b$):

| *Normal Set-up:* | *Co-axial Set-up:* |
|---|---|
| • 17 inches: (y) = $0.0016x^{1.1545}$ | • 17 inches (Top): (y) = $0.0014x^{1.2584}$ |
| • 18 inches: (y) = $0.0013x^{1.1888}$ | • 18 inches (Bottom): (y) = $0.0012x^{1.1892}$ |

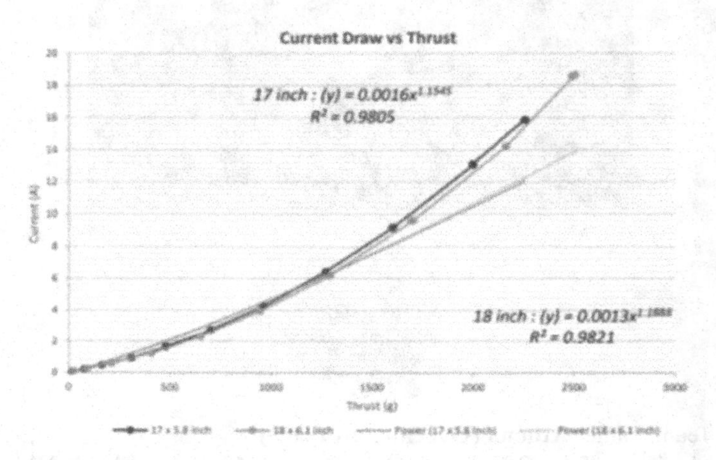


Overall one-off cost (£) (parts only): £ 1410
– Pixhawk FC.
Endurance (min) for 3 and 3.5 kg payload and 9000 mAh Multi-Star Li-Po battery:

| *Battery* | *Battery Mass (kg)* | *Payload (kg)* | *MTOM (kg)* | *Hover Endurance (min)* |
|---|---|---|---|---|
| 10,000 mAh 10 C | 1.189 | 3 | 6.64 | 4.5 |
| 8000 mAh 10 C | 0.956 | 3.5 | 6.907 | 3 |
| 8500 mAh 20 C | 1.020 | 3.5 | 6.971 | 4 |

## Guinness world record attempt 2018[2] (hover current record 2 h 6 m 7 s)

Team Name: **Enduro**
Maximum Take-Off Mass (kg) – without battery or payload: 0.97 kg
Number of motors: Four (Bespoke CF Arms hold 32 batteries)
Chosen Li-Po capacity (mAh) and mass (kg): (8S) 13,600 mAh and 1.462 kg (275 Wh/kg) (32 x Panasonic NCR18650B batteries)
Payload (kg): 0 kg
Motor Model Number and URL:

T-Motor U8 135 $K_v$ (42 POLES) http://store-en.tmotor.com/goods.php?id=323

Prop Size (Diameter x Pitch (inches)) and URL: 26″ × 8.5″ T-Motor http://store-en.tmotor.com/goods.php?id=407
Current draw per motor (A) @ hover (no payload, but with chosen battery): 1.71 A
Equation of Motor Current (A)(Y-Axis) Vs Thrust (g)(X-Axis) (in the form of Current = $aX^b$):

($0.0005X^{1.2573}$):

Show a graph of this curve:

2 www.guinnessworldrecords.com/world-records/longest-rc-model-multicopter-flight-(duration)

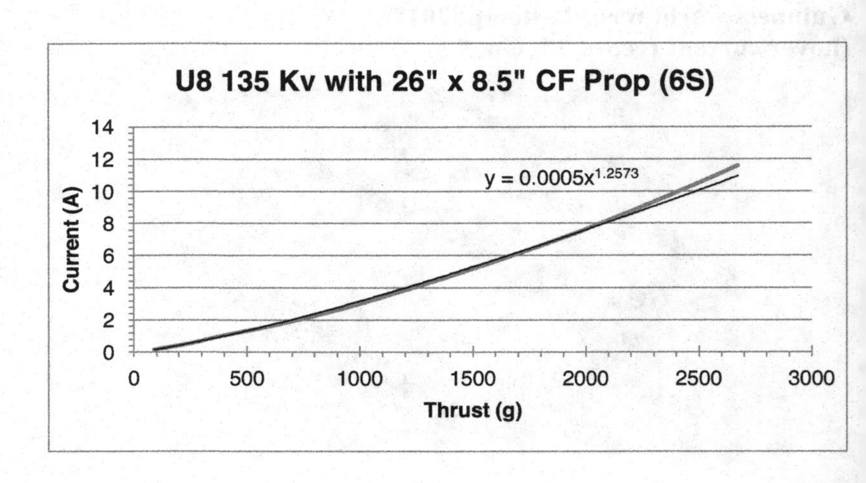


Overall one-off cost (£) (parts only): £ 3400

Endurance (min) for 0 kg payload and 13,600 mAh Li-Ion battery: 138 min

# Appendix F

## Drone radio frequency legal requirements in the UK

RF transmitters must comply with UK regulation set by Ofcom to avoid interference with other users.

> The use of the specific frequencies used for model control is exempt from the requirement to hold a Wireless Telegraphy Act 2006 providing that the apparatus meets the conditions of the IR2030.[1] In addition, the equipment must be lawfully CE marked and comply with all relevant EU directives. Apparatus that complies with the totality of the regulations does not require a licence to operate.

| *Frequency* | *Use* | *Effective radiated power* | *Comment* |
|---|---|---|---|
| 35 MHz | For telecommand to control the movement of airborne models only | 100 mW erp | 35 MHz band is dedicated solely to aeronautical modelling. EN 300 220 |
| 2.4 GHz | Any application, including wireless video cameras (non-broadcasting) | 100 mW with frequency hopping or 10 mW with other modulation | Commonly used for UAS. Reliable and low interference. Can also be used for airborne video transmission at 10 mW eirp. (2400–2483.5 MHz) EN 302 064 |

1 Ofcom (2018) *IR2030*. Available from: www.ofcom.org.uk/__data/assets/pdf_file/0028/84970/ir-2030-july-2017.pdf

| *Frequency* | *Use* | *Effective radiated power* | *Comment* |
|---|---|---|---|
| 434.04–434.79 MHz | For telemetry to provide data from the model only, including airborne models | 10 mW | Not exclusive to model controllers. Shared with other user with high power application. EN 300 220 |
| 458.5–459.5 MHz | For telecommand to control the movement of any model | 100 mW | Not exclusive to model controllers. Shared with other user with high power application. Certain long range devices make use of this. EN 300 220 |
| 5.8 GHz | Any application, including wireless video cameras (non-broadcasting) | 25 mW eirp | Can be used for airborne video transmission at 25 mW (5725–5875 MHz). EN 302 064 |

The most common type of frequency band used to control small UAS are 35 MHz and 2.4 GHz radio. However, 35 MHz radio is slowly being replaced by the more reliable 2.4 GHz radio. Since 2.4 GHz radio has a large bandwidth to accommodate frequency hopping modulation to avoid interference. Hence it is the ideal band for small UASs while complying the UK RF regulation. In addition, a 434 MHz data radio is commonly used for Data telemetry.

Note: As I understand the law, it is not illegal to sell RF TX equipment that exceeds the power output requirements of IR2030, but it is illegal to use it. The penalty is up to £ 5000 and/or six months' imprisonment.

# Appendix G

## Example Pixhawk flight controller wiring diagram and mission planner set-up (PID tuning)

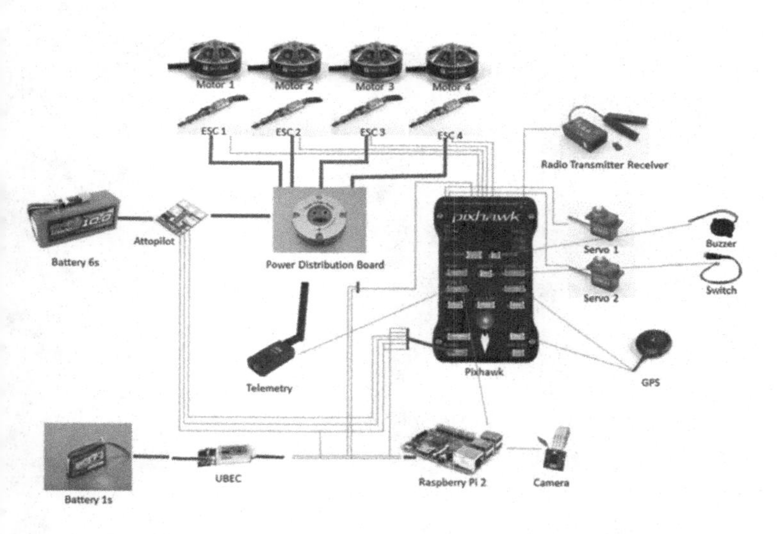


*Figure G.1* Pixhawk flight controller wiring diagram.

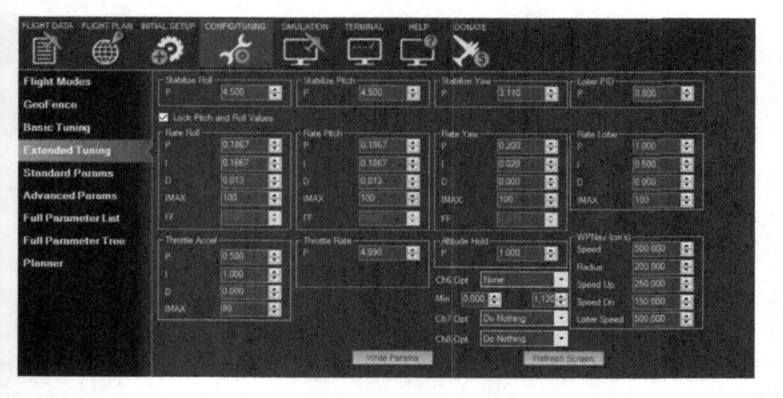


*Figure G.2* Mission Planner PID tuning set-up.

# Appendix C

## Example Pixhawk flight controller wiring diagram and mission planner set-up (PID tuning)

# References


Boeing Vertol Company (1977) *Heavy Lift Helicopter – Advanced Technology Component Program – Rotor Blade*. Available from: http://handle.dtic.mil/100.2/ADA053423

Brazinskas, M., Prior, S. & Scanlan, J. (2016) An empirical study of overlapping rotor interference for a small unmanned aircraft propulsion system. *Aerospace*, 3(4), 32. doi:10.3390/aerospace3040032

Bruce, E.S. (1903) Navigation of the air, "Juvenile lectures", lecture II. *RSA Journal*, 52, 165.

Corkery, J. (2016) *Actively canted co-axial quad*. Available from: www.chu.cam.ac.uk/news/2016/may/26/churchill-student-thrust-vector-drone/

Crowther, B., Lanzon, A., Maya-Gonzalez, M. & Langkamp, D. (2011, July) Kinematic analysis and control design for a nonplanar multirotor vehicle. *Journal of Guidance, Control, and Dynamics*, 34(4), 1157–1171, 14 p.

Drela, M. (2006, February) *Lab 3 Lecture Notes: DC Motor/Propeller Characterization*. Available from: https://ocw.mit.edu/courses/aeronautics-and-astronautics/16-01-unified-engineering-i-ii-iii-iv-fall-2005-spring-2006/systems-labs-06/spl3.pdf

eCalc (2018) *Multi-rotor Modelling Software Package*. Available from: www.ecalc.ch/xcoptercalc.php

Glauert, H. (1935) Airplane propellers. In *Aerodynamic theory* (pp. 169–360). Springer, Berlin, Heidelberg.

McMasters, J.H. & Henderson, M.L. (1980) Low-speed single-element airfoil synthesis. *Technical Soaring*, 6(2), 1–21.

Oxis Energy (2018) *Lithium-Sulphur Battery Technology*. Available from: https://oxisenergy.com/technology/

Prewitt, R.H. (1941). *U.S. Patent No. 2,234,196*. U.S. Patent and Trademark Office, Washington, DC.

Prior, S.D., Shen, S.-T., Erbil, M.A., Brazinskas, M. & Mielniczek, W. (2013) Winning the DARPA UAVForge challenge 2012 – Team HALO. *Unmanned Systems*, 1(2), 165–175. doi:10.1142/S2301385013400013

Ramasamy, M. (2015) Hover performance measurements toward understanding aerodynamic interference in coaxial, tandem, and tilt rotors. *Journal of the American Helicopter Society*, 60(3), 1–17.

Reyes, C. (2018) *Figuring Out Unknown Motor Specs*. Available from: http://rcadvisor.com/figuring-unknown-motor-specs

Srinivasan, V. (2008, September) Batteries for vehicular applications. *AIP Conference Proceedings*. AIP. Vol. 1044, No. 1, pp. 283–296.

Van der Merwe, C. (2018) *Delta vs Wye (Star) Termination*. Available from: www.bavaria-direct.co.za/info/

# Index

Printed in the United States
by Baker & Taylor Publisher Services